ENGINEERING
IN THE
CONFEDERATE HEARTLAND

ENGINEERING
IN THE
CONFEDERATE HEARTLAND

LARRY J. DANIEL

LOUISIANA STATE
UNIVERSITY PRESS
BATON ROUGE

Published by Louisiana State University Press
lsupress.org

Copyright © 2022 by Larry J. Daniel

All rights reserved. Except in the case of brief quotations used in articles or reviews, no part of this publication may be reproduced or transmitted in any format or by any means without written permission of Louisiana State University Press.

DESIGNER: Andrew Shurtz
TYPEFACES: Benton Modern, Clarendon MT, Scotch Roman MT

COVER ILLUSTRATION: Design for a pontoon train by James Norquet, 1862. Courtesy Jeremy Gilmer Papers, Southern Historical Collection.

Maps and diagrams by Mary Lee Eggart.

LIBRARY OF CONGRESS CATALOGING-IN-PUBLICATION DATA
Names: Daniel, Larry J., 1947– author.
Title: Engineering in the Confederate heartland / Larry J. Daniel.
Description: Baton Rouge : Louisiana State University Press, [2022] | Includes bibliographical references and index.
Identifiers: LCCN 2022005886 (print) | LCCN 2022005887 (ebook) | ISBN 978-0-8071-7785-3 (cloth) | ISBN 978-0-8071-7832-4 (pdf) | ISBN 978-0-8071-7831-7 (epub)
Subjects: LCSH: Engineers—Confederate States of America. | Engineering—Confederate States of America. | United States—History—Civil War, 1861–1865.
Classification: LCC E489 .D36 2022 (print) | LCC E489 (ebook) | DDC 973.7/3013—dc23/eng/20220222
LC record available at https://lccn.loc.gov/2022005886
LC ebook record available at https://lccn.loc.gov/2022005887

for Jennifer

Contents

Preface ix

1. Defending the Mississippi River 1
2. Scandal at the Twin Rivers 21
3. Engineering in the Field 38
4. Confronting Challenges 57
5. Engineering Colossus 71
6. Organizing Engineer Troops 93
7. The Mapmakers 112
8. We Want Engineers 121
9. The Pontoniers 139
   Epilogue 151

APPENDIX A: Engineer Officers of the Heartland 155
APPENDIX B: Occupations in the 3rd Engineers 159
Glossary 161
Notes 163
Bibliography 187
Index 197

Preface

"TRADITIONAL CIVIL WAR military historians need to stretch themselves out of comfortable niches imposed by their large popular audiences." Thus wrote Earl Hess in his essay "Revitalizing Traditional Military History." He argued that it was time to break out of the traditional molds of strategy, tactics, and generalship to explore new perspectives. Historians should now "cast our research nets far, wide, and deep." There are many "scantily explored" subjects, he argued, including how the various behind-the-scenes departments operated. The following pages attempt to address one of those unheralded departments: the engineers.[1]

James L. Nichols's 1957 small introductory study of the Confederate engineer corps curiously remains to this day the central book on the subject. By contrast, several superb books have been published on the Federal engineers. Rather than writing an expansive volume on Confederate engineering in the Army of Northern Virginia and along the coastal areas, both of which are worthy topics, I have determined to focus on the "Heartland." It was Thomas L. Connelly who, in 1967, first coined the term "Heartland." He was referring to the region of Tennessee, north-central Alabama, north-central Georgia, and northeast Mississippi, the operational area of the Army of Tennessee. While accepting the name, which is commonly used by historians today, I have expanded upon Connelly's definition by including military operations at Vicksburg, Mississippi, a portion of Louisiana, and East Tennessee. All of these sectors were connected; indeed, Mississippi and East Tennessee served as the strategic flanks of the Army of Tennessee. Additionally, the engineers of the Heartland frequently traveled back and forth in these sectors. It is difficult to imagine the sheer magnitude that the Heartland embraced—no less than 175,000 square miles. It represented the equivalent size of modern-day Germany, Belgium, and Switzerland combined!

Complicating the work of the engineers was the wide geographical diversity of the Heartland. The immense area extended from the Unaka Mountain section of the Appalachian Mountains (with elevations of 5,000–6,600 feet) to the Cumberland Plateau (with elevations of 400 feet and sandstone bluffs cut by several northeasterly running valleys) to the low-lying Nashville basin's thick cedar glades and rolling farmland to the wind-blown silt soil of the Loess Plains of the Mississippi Valley. Indeed, if there was any unifying factor in the geographic Heartland, it was flooding, which occurred in all sectors. The Upper Mississippi River gauge in 1848 was 26 inches, 49 inches in 1850, and, during the Great Flood of 1851, 74.5-inches in a three-month period! Flooding in 1862 along the Tennessee portion of the Mississippi River was described as "epic."[2]

George G. Kundahl's excellent book on the life of John Morris Wampler was the first glimpse behind the curtain of life in the western engineer corps. Unfortunately, Wampler was in the Army of Tennessee only fifteen months (April 1862–July 1863), and he missed both the Battle of Shiloh and, due to illness, the Battle of Stones River. This volume takes a view of the entire Heartland area throughout the war. I once had a retired army general smilingly tell me: "Amateurs study battles; professionals study logistics." Armies are indeed complicated institutions, in which no component can fail. The small western engineer corps was as vital an element to implementing strategy and tactics as arming, feeding, and equipping the armies.

The Heartland did not operate independently, but within a national system of engineering, under the direct command of the Engineer Department in Richmond, Virginia. Initially the Confederate Congress authorized a tiny bureau comprised of a colonel, a handful of officers, and a single company of sappers and miners. Following a formal declaration of war, the department was increased, but only marginally, with the addition of a few more officers and another company of sappers and miners. For a number of months the department operated under the leadership of a succession of "acting chiefs," including W. H. C. Whiting, Josiah Gorgas (who doubled in his role as chief of ordnance), and Danville Leadbetter. Thirty-one-year-old Alfred L. Rives, a highly trained engineer, served through the summer of 1862. Eventually Colonel Jeremy Gilmer assumed command in September 1862, despite his desire to remain in the field as chief engineer of the Army of Northern Virginia. The bureau offered support in supplying tools and specialized equipment, reviewed contracts, and dealt with budgets and the constant requests for promotions. Overwhelmed and short-handed, the work of the Engineer Department often had mixed results in the Heartland.[3]

Hopefully the technical issues of flood plains, right angles, hachures, and elevations did not get in the way of the larger story of the men, most of whom had no military training, but who were called upon to transition from the civilian tasks of water drainage, railroad construction, and land surveys to perform highly technical military engineering skills, and with scant time for a learning curve. A few of the engineers were academics, such as Andrew Buchanan, who taught engineering at private colleges, but even in those cases there is no indication that they had practical experience outside the classroom. The same was true of West Point graduates, who, though they passed through Dennis Hart Mahan's demanding engineering classes at the academy and received limited training in field fortifications and defenses, upon graduation were then typically assigned to coastal forts or frontier posts or, after a time, quit the army for civilian pursuits. Several of the West Point-trained engineer officers who served in the Heartland, such as Samuel H. Lockett, Joseph Dixon, David B. Harris, and James M. Couper, had previously only served in a peacetime army. Thus, even for the military professionals, there was often a gap between theory and practicality. Like country doctors who had to quickly learn how treat grisly battlefield wounds, or foundrymen who were called upon to cast cannon and projectiles rather than saws and ploughs, the engineers had to evolve.

Many books have been written about the battles and campaigns of the Army of Tennessee and the Siege of Vicksburg, and to some minor extent the role of Confederate engineering has been touched upon. An in-depth examination of the personalities and backgrounds of the engineers, and a comprehensive examination of what they got right and where they failed, had yet to be written. There were surprises that emerged. First, the amount of talent in the Heartland belies the traditional thesis that the area was void of professionals. Second, while civil engineers were frequently looked down upon by military professionals, the most egregious engineer mistakes were usually made by West Point-trained officers. Third, though the concept of engineer troops, as opposed to the professionally trained officers, was slow to come of age, their skills and labor, especially in constructing small bridges and the laying of pontoons, often proved pivotal. Fourth, by 1864 the Confederates in the Heartland had mastered map photography, a fact not generally appreciated.

The goal of Confederate engineers in the Heartland was simple in mission but complicated in implementation: to use their specialized skill sets to defeat or slow the Union juggernaut. Despite their vital contribution to the war effort, their names have received but scant recognition in official reports, if mentioned at all. My goal was to bring these men and their accomplishments out of their obscurity to the extent that sources permitted. To enhance the narrative, I also spent a great deal of time gathering a number of previously unpublished photographs, illustrations, and maps, in an attempt to avoid boring the reader with stock photographs that have been used in countless general histories.

PREFACE

If there were any branches of the service in which the Federals excelled, it would be the artillery and the engineers. The Federal engineer corps in the West certainly surpassed their counterparts in organization, technical sophistication, and the size of their talent pool. Remarkable engineering projects, such as the canal around Island No. 10, the speed with which railroad trestles could be rebuilt, the professionalism of the military maps that were produced, the early organization of engineer troops, and even the ingenious idea of using cotton bales as pontoons in the Vicksburg Campaign, were not surpassed by the Confederates. The southerners were hindered by a lack of the sophisticated equipment and technical instruments so vital to engineers. Even so, the Confederate engineers had their own successes—neither the Vicksburg nor Atlanta defenses were ever breached, and by late 1864 the Army of Tennessee boasted a pontoon train sufficient to span the Tennessee River. Admittedly, large-scale bridge projects could not be completed at the speed of Federal engineers. Bluecoat engineers rebuilt a railroad trestle over the Chattahoochee River in six weeks, whereas it required the Confederates two and a half months to reconstruct a bridge of similar length over the Holston River. Large-scale projects nonetheless were completed, albeit behind schedule, notably at Bridgeport, Alabama, and with several large bridges in East Tennessee.

Some technical terms are inevitable in a book of this sort, and a glossary is included to assist in definitions. I wrote much of this book during the 2020 pandemic, when archives, libraries, and parks were essentially closed. Without the assistance of archival staffs, this project could not have been completed. I am particularly appreciative of Susan Hawkins of the Fort Donelson National Battlefield Park, Terry Winschel (retired) of the Vicksburg National Military Park, and Jim Ogden of the Chickamauga and Chattanooga National Battlefield Park. I also am indebted to Benjamin F. Cooling, who gladly shared his journal by John Haydon, which added another perspective to the Twin Rivers chapter. Stephen Davis of Atlanta offered invaluable information about the Bushrod Frobel diary and where it could be located. He also enthusiastically shared additional notes. Michael Shaffer advised me on information regarding the Chattahoochee River Line. Timothy Smith, a long-time friend, assisted in the Vicksburg chapter. Most of all, I express thanks to Justin Solonick of Grapevine, Texas, Sam Elliott of Chattanooga, and Dave Powell of Chicago, all experienced authors, for reading some or all of the chapters and offering their corrections and suggestions for strengthening the manuscript.

ENGINEERING
IN THE
CONFEDERATE HEARTLAND

1
DEFENDING THE MISSISSIPPI RIVER

THE WESTERN THEATER, the so-called Confederate Heartland, comprised the sprawling territory between the Appalachians and the Ozarks. During the first two years of the Civil War it was the rivers in that area, not the railroads, that proved pivotal. Foremost on the minds of both Federal and Confederate authorities was the Mississippi River, characterized by Steven Woodworth as "both the great east-west divider of the continent and the great north-south conduit of commerce." Northwesterners trembled at the potential commercial loss that would result if free navigation was interrupted. Southerners likewise quaked at the strategic disaster that would occur if the Confederacy was dismembered and vital cities captured. Fear ruled, and both sides began developing strategic initiatives to control the river.[1]

Captain Justus Scheibert, an engineer, teacher, and up-and-coming star in the Prussian army, traveled to the Confederacy for seven months in 1863 to observe what could be learned from the war. Although he never came to the West, he studied reports and maps, both Federal and Confederate, and arrived at several conclusions. He observed that though the tributaries of the Mississippi River were not as deep as the rivers of central Europe, they nonetheless permitted light-draft steamers and gunboats to ascend, thus "The side that held the Mississippi would hold the strategic advantage."[2]

The North soon began construction of a fleet of ironclad vessels to conduct offensive operations. The South also began laying the keel of two ironclads in Memphis, but the work proved tediously slow. It was clear that the primary defense of the rivers would come from fixed earthen fortifications. The surveying and construction of such forts would require military engineers, and they were in short supply. The South, therefore, had no choice but to turn to a handful of civil engineers, architects, and surveyors. Their valiant, though at times faulty and ineffective, military engineering attempts temporarily hindered the Union's ability to seize key rivers necessary for their southern invasion.[3]

Early Fortifications

In April 1861, the pretentious and petty Major General Gideon Pillow assumed command of the Tennessee Provisional Army. Although a veteran of the Mexican War, he had no formal military training. He nonetheless was an ardent secessionist, held an association with Governor Isham G. Harris, and possessed great wealth. As a nonprofessional political general, Pillow lacked the modesty to see his own weaknesses. Confederate secretary of war Leroy P. Walker advised state authorities to prepare a battery on the Mississippi River at Memphis and make preparations for additional fortifications upriver. He believed that the Federals would certainly plan for an early river descent with the purpose of burning cities along the way. Rumors floated of large Federal concentrations at Cairo, Illinois. In an article entitled "Our River Defenses," the *Memphis Appeal* reassured its readers that the city's security was well in hand; nothing was further from the truth.[4]

The subject of engineers topped the general's concerns. "Can you send an engineer here [Memphis] capable of directing military defensive works?" Pillow wrote Walker on April 20. The secretary replied that "an engineer" would soon arrive. Twenty-nine-year-old Lieutenant Philip Stockton, a West Point graduate, reported for duty in late April. At the beginning of the war, West Point was seen as the nation's premiere engineering school, with a curriculum heavily weighted toward engineering. During their senior year, cadets studied under the widely respected (although not universally popular) professor, engineer, and author Dennis Hart Mahan, referred to by Justin Solonick as the "Academy's engineering guru." Only the top graduates went into the elite US Corps of Engineers or US Corps of Topographical Engineers. Classroom theory and instruction notwithstanding, Stockton, along with most West Pointers, lacked extensive post-graduate experience. Indeed, in the Regular Army Stockton had fought Indians and he currently served as an artillery officer. The construction of large earthen fortifications would require a steep learning curve.[5]

To meet the increasing engineer demands, Pillow turned to a cadre of Memphis civil engineers. Thirty-four-year-old William D. Pickett was born in Huntsville, Alabama, but raised in Kentucky. He was in Texas conducting a survey for a land dispute when the Mexican War broke out. He promptly joined the Texas Rangers and fought Comanches. At the start of the current war, he served as civil engineer for the Mobile & Ohio Railroad with an office in Memphis, Tennessee. He organized a company of sappers and miners, along with thirty-four-year-old Minor Meriwether, chief engineer of river levees in West Tennessee, twenty-six-year-old Edmund W. Rucker, a railroad civil engineer, forty-three-year-olds Calvin Fay and Deitrich Wintter, both Memphis architects, Joseph A. Miller, a draftsman and surveyor, and Edward McMahon, a machinist. Pickett subsequently received a captain's commission in the state army and was transferred to Pillow's staff, leaving Wintter in command. Sappers and miners were technically formed to work on fortifications, but as time went on and field demands increased, they would become the forerunners to engineer troops.[6]

Work began in Memphis at the site of old Fort Pickering on the third Chickasaw Bluff. Built by the French in 1739 on top of existing Indian mounds, the US Army used it as a trading post from 1798 to 1814. Nearly a half-century of neglect required intense labor. To bolster defenses, cotton bales were placed along the entire city front. "The river defenses, which have gone up almost like magic under the auspices of General Pillow and his engineering corps, are sufficient to sink any fleet of gun boats," the *Appeal* boldly, and quite naively, declared. A visiting London correspondent later viewed the works and expressed astonishment at their weakness. The fort was located so close to the edge of the bluff that it would "offer no cover against vertical fire, and is so placed that well-directed shell into the bank below it would tumble it into the water," wrote William Russell. "The zig-zag roads are barricaded with weak planks, which would be shivered to pieces by gun-boats; and the assaulting parties could easily mount through these covered ways to the rear of the parapet, and up to the very center of the esplanade." The work had been given to novices, with predictable faulty results.[7]

Meanwhile, Meriwether and his sapper company began work on Fort Harris (named after Governor Harris), six miles above Memphis at a bend in the Mississippi known locally as "Paddy's Hen" or "the Chickens." The small earthwork had a sixty-four-foot base and an eight-foot-high parapet with an interior slope of two and a half feet, and an elevation twenty-five feet above ordinary stage. The old tactic of stretching a chain across the river was attempted. In 1588 Queen Elizabeth I had ordered a chain stretched across the Thames River to prevent the ascent of the Spanish Armada. George Washington did the same thing during the American Revolution to protect the Hudson River. The thick timber to the north and west of Fort Harris was cut down to form an abatis. Initially 280 hands performed the labor, mostly slaves supplied by local planters, but pleas were made for an additional 200 Blacks to clear brush. The weak fortification was never designed to be much more than a battery.[8]

Pillow pleaded with the Confederate authorities for more engineers, but Walker had none to send. A Tennessee state engineer, fifty-one-year-old, Virginia-born Captain Montgomery Lynch, a former civil engineer of the Memphis & Little Rock Railroad, did arrive, as did James H. Humphreys, a civil engineer for the city of Memphis. An additional twenty-seven slaves bolstered the workforce.[9]

While work at Memphis and Fort Harris continued, Pillow labored over tactical issues. Relying on Federal recognition of Kentucky neutrality, Tennessee authorities suspected that the bluecoats would march down the west bank of the Mississippi River through Missouri or Arkansas and attempt an amphibious landing. Once on the east bank, the Federals could take river fortifications in reverse. Pillow thus devised a stacked formation defense with a string of fortifications that ran north to south in an attempt to both defeat enemy gunboats and cover all potential landing sites. This created two problems. First, with no defined east-west defensive line, the Federals, if Kentucky neutrality was ever nullified, could simply march through West Tennessee to Memphis, bypassing all of the forts. Second, a tiered river defense meant that slave labor, heavy artillery, supplies, and engineers would be diffused rather than concentrated.[10]

The next fortification, Fort Wright at Randolph, located sixty-five miles north of Memphis and immediately south of Island No. 34, presented a far greater engineering challenge. Stockton, Lynch, Humphreys, Pickett, and Wintters's sapper company all worked on the project. A northern correspondent later described the town as a "half dozen or so dilapidated houses." There was no "fort," per say, but a maze of rifle pits. The heavy artillery platforms were located behind twenty- to thirty-foot-thick earthworks and arranged in three tiers—"somewhat like an ancient amphitheater, rising gradually to the summit of the bluffs." Enemy boats would have to pass within 400 yards of the terraces under a plunging fire. The works were sodded with grass to blend in with the rugged terrain. In the rear of the fortifications, a fifty-foot-high sand bluff spelled doom for the defenders if occupied by the Federals, but a Memphis reporter offered assurances that the position could not be turned. There was a narrow defile on the land side that was covered by a lunette. When William Russell examined the defenses in June, however, he remained unimpressed. The batteries were "rudely erected in an ineffective position," and the gunboats, he believed, could easily pass the heavy artillery, and as the river fell the guns would be even less effective. To slow a rapid passage of enemy gunboats, Pillow had a ship-cable chain, supported by rafts and buoys, stretched across the Mississippi River.[11]

An Arkansas regiment, under Colonel Patrick Cleburne, an Irishman who had served in the British army, was sent upriver from Randolph to the eighty-foot-high first Chickasaw Bluff and established Fort Cleburne. A sandbar protruded in the middle of the river, with the main channel between it and the eastern shore. Pickett, Lynch, Wintter's sappers, twenty-seven Irish laborers, and forty-two Blacks were sent to start construction on a massive fortification, subsequently renamed Fort Pillow. On the land side, a zig-zag semi-circular line of breastworks, capable of accommodating 20,000 troops and fronted by a ditch and felled timber, stretched from Coal Creek to the north and the Mississippi River to the south. A northern correspondent later expressed amazement at the "scientifically made" outer defensive parapets, lined with timber and secured by deeply driven posts. The water batteries, at the base of the bluff, stretched for half a mile, with bastions built of sandbags, planking, and rammed clay. Lynch devised a plan (condemned by some) to place pilings in the river to narrow the channel, thus slowing the gunboats. Although the laborers were novices at such work, they eventually got "the hang of it."[12]

In July 1861 Major General Leonidas Polk, a West Point graduate, Episcopalian bishop, and long-time associate of President Jefferson Davis, assumed command of Department No. 2, which included the Mississippi Valley. The Tennessee state forces were merged into the Confederate army, leaving Pillow with the reduced rank of brigadier. Indignant and resentful, Pillow soon clashed with "The Bishop," as he disrespectfully referred to him behind his back. Polk authorized a 6,000-man "Army of Liberation," under Pillow's command, dispatched to New Madrid, Missouri. The Tennessean's pipe dream of an offensive to Cape Girardeau and ultimately Saint Louis, Missouri, soon fell apart, due largely to sickness in the ranks and exaggerated numbers. Polk would go no farther, leaving Pillow stuck in New Madrid and bickering with his superior, at times rising to the level of insubordination.[13]

In August, forty-one-year-old, Virginia-born Captain Andrew B. Gray, Polk's topographical engineer, arrived at Island No. 10. Early in his career he completed an apprenticeship in surveying and engineering, but there is no evidence that he had formal training in the latter. In 1842 he drew some maps of the mineral lands adjacent to Lake Superior, but he was best known for surveying the boundary between Texas and Mexico following the Mexican War and, in 1849, surveying and laying out the town of San Diego, California. In 1854 he did extensive surveying for the Texas & Western Railroad; two years later he relocated to New Orleans. Gray's appointment to Island No. 10 would be problematic. The job required a civil engineer, not a surveyor, but the talent that summer was spread thin.[14]

Fig. 1.1. Captain Andrew B. Gray. It is not known why Gray, the engineer in charge at Island No. 10, signed his name as "Asa B. Gray." Courtesy San Diego History Center.

Island No. 10, one mile long and ten feet above low water, was located in an upside-down U-shaped bend in the Mississippi River just south of the Kentucky border. Captain Scheibert described it as "two horseshoe bends, New Madrid in one, Island Number 10 in the other, and both in marshy terrain." The primary river channel was between the island and the Missouri shore, but in high water the Tennessee chute was also navigable. Six miles to the south was Tiptonville, Tennessee, later characterized by a northern correspondent as "a miserable collection of log shanties." The peninsula was heavily timbered, and the shore along the Tennessee bank frequently flooded. On the Missouri shore, north of the island, was New Madrid, "the weak point" according to the *New Orleans Delta*. Steamboats had to pass the town before they reached the island. If the Federals captured the town, river traffic to the island would be severed.[15]

Gray made a "rapid reconnaissance in a few hours" and work began on August 19 on Battery No. 1, named Fort Polk, a redan on the Tennessee shore a mile north of the island. It was never expected to be held in extreme high water, Gray later claimed, but to "produce a certain effect for a certain period." There were sufficient wheelbarrows, but this proved useless for the time being because of a lack of shovels, which had to be sent from Fort Pillow. When Pillow arrived for an inspection at the end of the month, he expressed disgust. The redan had been poorly located in a muddy area subject to a three-foot overflow;

indeed, the river was already eroding the west wall. As for the interior of the fortification, "the engineer [Gray] is now grading down the original bank within the work fully two feet, so that the seep water will drive out the forces in the work before the river gets within three or four feet of high-water mark." Gray complained that he had been forced to make embrasures (gun ports) on the redan that weakened the parapet, since three of the guns had been mounted on sliding naval carriages made in New Orleans. He preferred that the heavy guns be mounted on "high barbette" carriages, or peering over the parapet.[16]

Pillow's criticism notwithstanding, Gray continued construction on the redan, where four guns had already been mounted. Although no obvious topographical advantages existed, he insisted that the position was a strong one. There would have to be a 1,100-foot cremailliere line constructed between the redan and a five-foot-deep bayou to guard the northern approach. Reelfoot Lake, which at places was eleven miles wide, protected the eastern land approach. The main concern was the shortage of labor. Gray heard that 2,000 slaves were engaged at Fort Pillow, and he insisted that he needed 500 of them to complete the defenses. Adding to his woes, sickness had stricken his engineer assistant, Lieutenant J. Hudson Snowden. Although listed as a topographical engineer, it appears he was an infantryman detailed "as an engineer." A greater loss was that of Lieutenant Robert P. Rowley, the twenty-three-year-old former civil engineer for the state of Arkansas. He had joined the 4th Tennessee Infantry Regiment at Memphis, but Pillow detailed him to the engineers. He briefly worked at Island No. 10, but he was reassigned to East Tennessee in September.[17]

By early September 1861, despite massive exertion, little progress had been made on the Mississippi River fortifications. Memphis had two batteries with a total of ten heavy guns, Fort Harris had no guns mounted, Fort Wright at Randolph had four batteries with eighteen guns, Fort Pillow had twelve guns, and Island No. 10 had one battery of four guns. The primary problem was that the overwhelming majority of the pieces were smoothbore 32-pounders. The effectiveness of such guns in direct frontal fire against Federal ironclads, known to be under construction, would be minimal.[18]

Columbus

Pillow now saw the opportunity to revive his long-held dream of occupying Columbus, Kentucky, on the Mississippi River twenty miles from the Tennessee border. The site was far superior to Island No. 10, or so he thought, but of course there was the sticky issue of Kentucky neutrality. The Kentucky state legislature, hoping to avoid getting caught in the middle, had not declared for either side. Pillow, with only slight urging, nonetheless convinced Polk to occupy the town. On September 4, without notifying the War Department, the bishop-general ordered the town seized. The operation, though popular with many, presented immediate and serious problems. The Federals, under General Ulysses S. Grant, quickly moved to occupy Paducah. The 400-mile-long Tennessee-Kentucky border, which the Confederates were totally unprepared to defend, now lay ripe for invasion. Although an astonished Davis initially opposed the move, he ultimately acquiesced—the damage had been done.[19]

Throughout September, nearly 12,000 troops (their "clothes seemed old and some even ragged," observed one) flooded into Columbus. Lacking sufficient slaves to perform the labor on the fortifications, the soldiers were placed in fatigue details. A Tennessean "looked at the different Batteries and breast works[.] [M]en were at work as for life and death making land Batteries while upon the heights above the Town the smoke of the enemies boats was plain to be seen over the tops of the timber some 10 or 15 miles off." Work continued on the heavy batteries. Edmund W. Rucker reported on September 22 that the battery "under the cliff" was at a ninety-seven-foot elevation level at low water mark, and at high water sixty-four and a half feet. Initially Polk had viewed Island No. 10 as the advance position on the Mississippi River, with Fort Pillow as the backup position. Columbus now became "the most important point," with Island No. 10 as the reserve. Fort Pillow would be "the last stronghold in the chain."[20]

In mid-September 1861, the highly anticipated General Albert Sidney Johnston, the newly appointed commander of an expanded Department No. 2, arrived in Nashville. An artilleryman described him as a "medium sized portly gentleman of about forty-five and upwards [fifty-eight] & dresses in citizens clothes & slouched hat & upon first sight would not impress a stranger favorably as to be anything more than an ordinary man." Johnston was not ordinary, but he was, as matters developed, overrated. He immediately faced a decision—either withdraw Polk's Corps or advance his center corps under Major General William J. Hardee to Bowling Green, Kentucky; he chose the latter. On the 18th, Johnston arrived at Columbus for an inspection. The visit was as much a social reunion between he and Polk as it was a strategy session. Indeed, every indication is that Johnston embraced Polk's ill-advised decision to occupy the town.[21]

With so many fortifications along the Mississippi River under simultaneous construction, the engineer corps was stretched beyond capacity. Johnston immediately informed Adjutant General Samuel Cooper of the dearth. "The necessity of engineers is pressed on my attention by the wants of every hour. Can they be furnished? If not, can I muster the engineers of Tennessee [Provisional Army], if to be had?" Replying that same day, Copper answered: "I have tried to find you an officer, but have failed. There are but few in the [C.S. Engineer] corps, and they are on important duties. You must have several [West Point] graduates in your command, some of whom will answer the purpose." In his follow-up dispatch the next day, September 26, Cooper advised Johnston: "Do the best you can in respect to engineers. Employ any officers you can find." Secretary of War Judah Benjamin sounded a similar note, stating that there were but six captains and three majors in the Regular Confederate army, and one of the last was employed on bureau duty. "You will be compelled to employ the best material within your reach by detaching officers from others corps and by employing civil engineers, for whom pay will be allowed."[22]

Pickett was ordered to Columbus, but getting additional engineers would continue to be problematic. Matters improved, but only briefly. Stockton was detailed as Polk's inspector general in August. Replacing him would be sixty-six-year-old Major Lewis DeRussy. A West Point graduate, DeRussy served for twenty-nine years in the Regular Army variously employed as an engineer, surveyor, and topographical engineer. He subsequently led a Louisiana regiment during the Mexican War. At the time of the present conflict, he worked as a Louisiana civil engineer. As Polk's chief engineer, DeRussy worked at Columbus throughout the fall, but was soon transferred to Louisiana. Polk detailed Lieutenant Joseph A. Miller from the artillery to the engineers. Although not a civil engineer, Miller had previously owned a foundry in Memphis and was something of a mechanical genius. He devised a superior artillery carriage and projected a plan to develop a torpedo (mine) to place in the river channel. He was ordered to oversee the construction of bombproofs for the two lower batteries at Columbus, and for that he was given 150 axes, but no saws or wagons; the project quickly stalled. Miller warned that once the rainy season began it would be impossible to complete the work.[23]

Some officers began to question the vulnerability of the Columbus defenses. An ordnance officer believed that Union ironclads could easily slip past the bluff batteries one dark night. "In the excitement of the moment, your gunners would not hit them once in a hundred shots," wrote J. T. Trezevant. Colonel Marsh Walker, a West Pointer, came to a similar conclusion. A civilian made the observation that too many guns were atop the bluff and that more guns were needed "for a horizontal fire, which, of course, is best for inexperienced gunners. The plunging fire from the high bluffs is admirably fitted for the sloping sides of the gunboats, but will be quite uncertain."[24]

By mid-October, construction on the most prominent bluff work (Fort DeRussy) was nearing completion—and none too soon. On November 6, Federal troops under Grant attempted an amphibious landing to seize the Confederate camps across the river from Columbus at Belmont, Missouri. In conjunction with the movement, the wooden gunboats *Tyler* and *Lexington* made a feint against the Columbus batteries. The bluecoats were repulsed at Belmont and the gunboats withdrew, *Tyler* having sustained a serious hit. Four days later, while test firing a heavy artillery piece, an explosion occurred, wounding Polk, killing Lieutenant Snowden and others, and tossing Captain Pickett fifty feet, incapacitating him for nearly a week and permanently impairing his hearing. Rucker was appointed to the artillery that month—another engineer gone.[25]

With DeRussy's transfer, Minor Meriwether was promoted to major as Polk's chief engineer. By the end of December 1861, he had five engineers working the Columbus fortifications.[26]

Eventually Columbus became a massive fortress, with five large earthen forts and miles of trenches extending on either side of the Clinton Road. The engineers decided on a tiered defense, with a large river battery (twenty-two guns) fifteen feet above the river's edge, mid-level batteries on the bluff at about eighty-five feet, and a third level atop the bluff at 180 feet. Although the main work was christened Fort DeRussy, local historians today claim that another engineer (presumably Pickett) did most of the work. DeRussy was present during the construction, however, and almost certainly he was in charge. The land defenses extended along the bluff heights for nearly two miles, with field redoubts seven to eight feet wide. The most vulnerable of the river redoubts (No. 4), at the foot of the bluff, being in a direct line of fire, had a parapet thickness of nineteen feet. A one-mile-long chain with twenty-pound links stretched across the Mississippi. The troops encamped atop the bluff, with water obtained by means of a steam pump that emptied into a large tank. A northern correspondent later pronounced the defenses to be "in the most thorough manner, upon purely scientific principles, showing the handiwork of a master engineer."[27]

Work meanwhile continued at Fort Pillow under Montgomery Lynch. On October 19 he reported to Polk that 1,400 Blacks were at work and the project was progressing, although "there is yet much work to be done." By the end of November the situation had turned for the worse. By that time only forty-two slaves remained, some of the labor having been sent to New Madrid and Island No. 10. "I am just now down in spirit to the completion of the work here," Lynch notified Polk on November 26. If the project was not completed in time, he fretted that the blame would be attributable "to the ignorance of

Table 1.1. Engineers at Columbus, December 1861

ENGINEER	ASSIGNMENT
Minor Meriwether	Chief Engineer, Columbus
Joseph A. Miller	Mounting guns and building magazines at base of bluff
Arthur W. Gloster	Constructing three magazines on spurs of hills
Henry N. Pharr	Placing submarine batteries (mines)
John C. Mann	Constructing quarters and traverses in Fort DeRussy
William D. Pickett	Completing earthworks where railroad enters the bottoms

a civil engineer." Captain Wintter's sapper and miner company had not been paid in three and a half months. There was talk (wrongly) that the company would be converted to artillery, leaving him with no mechanics. Despite his pleas, Wintter's company was ordered to Island No. 10. On December 1, Lynch informed Polk: "We have platforms ready for sixteen guns and room for eighteen more platforms not yet built."[28]

Spies were also taking notes. A report was published in the *St. Louis Republican* on November 7 by a "Union man" proceeding downriver. At Columbus he noticed two batteries of eight or ten guns each, but he noted that work was progressing day and night on the batteries above the town. At Island No. 10 he saw "a masked battery," apparently the redan, but he could not tell the number of guns. At Fort Pillow, at the site of Bragg's sawmill, anchors and chain cable had been unloaded. Fort Harris had eight guns but not over fifteen men. At the mouth of the Wolff River he observed a battery of five or six 32-pounders, and at the foot of Jefferson Street in Memphis he saw a battery of 32s.[29]

On February 4, 1862, General P. G. T. Beauregard reported to Johnston at his Bowling Green, Kentucky, headquarters as his new second in command. Politics was clearly behind Davis's decision to transfer (some have suggested exile) the Louisianan to the West. The *Boston Journal* wrote of the West Pointer and former Old Army engineer that "if Beauregard is good for anything, it is as an engineer," while the *New York World* pronounced him "the most marvelous engineer in modern times." The *Appeal* declared that Robert E. Lee and Beauregard were the "most competent engineers on the American continent." The Louisiana general was dismayed at what he found. Johnston's two corps at Columbus and Bowling Green were in advance of his center at Forts Henry and Donelson (to be discussed). The deployment, according to his biographer, "shocked his every engineering sensibility."[30]

Beauregard realized that Columbus could be invested or even bypassed.

On February 19, he dispatched his chief engineer, Captain David B. Harris, to inspect the town's fortifications. Harris's receding hairline and spectacles made him appear older than his forty-seven years. He had served in the artillery upon his graduation from West Point, but he subsequently taught engineering at the academy until his resignation in 1835. In civilian life he became a Virginia tobacco farmer and exporter. At the outbreak of the war, Harris initially served in the Virginia Provisional Army, until it merged with the Confederate army. His subsequent evaluation of the Columbus defenses proved damning. The land-side works were "defectively" located, "badly planned and unfinished," and "much too extensive." There were guns with no traverses and wooden warehouses liable to catch fire. Unable to personally travel due to sickness, Beauregard summoned Polk to his new headquarters at Jackson, Tennessee. Union forces, Beauregard argued, could march west from Fort Henry to Clinton and invest Columbus from the west or to Union City, Tennessee, and take Columbus in reverse. The "Gibraltar of the South," as it had been called, had to be abandoned and the guns sent to Island No. 10. Polk, failing to see that Columbus was a trap, argued that the land defenses could be strengthened. The decision (subsequently approved by the War Department) had nonetheless been made.[31]

Underlying the Columbus evacuation was a serious divide between Polk and Beauregard that ultimately affected the task assigned to engineers. Polk believed in an all-out river defense, while Beauregard, understanding the relationship between engineering and operational tactics, subscribed to a modified defense. He believed that armies should be kept mobile. Captain Harris estimated that it would require 13,000 troops to adequately man the Columbus defenses, whereas Beauregard had 3,500 in mind. The same issue would arise at Fort Pillow. It was built for a garrison of 15,000–20,000, but the Louisianan desired an inner line for no more than 3,000. "[D]efensive lines must not be too extensive," Beauregard harped. He desired an "entirely different system," whereby the Mississippi River fortifications would be "almost entirely reconstructed for minimal garrisons." On the face of it, Johnston seemed to agree. He had earlier stated that if Columbus was invested, Polk's "army" could come to the rescue. The theater commander appeared oblivious as to the sheer magnitude of the Columbus fortifications, which would have required most of Polk's Corps as a garrison. If the city was invested, there would be no Confederate army coming to break the siege.[32]

General Joseph E. Johnston, who would later become the Western Theater commander, adhered to Beauregard's strategy. He opposed the fortification sizes at Columbus, Island No. 10, Forts Henry and Donelson, and Bowling

Green, Kentucky, "each of which required an army to hold it; and, consequently, a respectable army divided among them, gave each one a force utterly inadequate to its defense." He believed that garrison strengths should be 1,000–2,000 men, saving enormous labor and adding strength to the larger "armies" (actually only two corps). "As it was, the Confederates were alike weak at every point," he concluded. There is no indication that he actually expressed such a view at the time, but only years later as he was writing his memoirs.[33]

The obvious flaw to the Beauregard-Joseph E. Johnston strategy of maintaining small garrisons and sending reinforcements only when they were threatened was that it assumed that the "parent" army could break away from its own operations to send sizable reinforcements. Such a plan, although theoretically sound on paper, was almost certainly implausible if the Federals either coordinated their movements or advanced in overwhelming strength. There was another possibility that, although more realistic, would have been politically unacceptable. The Confederates could have conceded that the forts were mere stopgaps meant to buy time and only marginally defend the rivers. The only possible major victory to be won was once the Federal armies advanced inland away from the protection of their ironclad fleets. Such a strategy would, of course, have been rejected by the Davis administration, given the political ramifications. The point is, strategy determined engineering goals. The Beauregard-Johnston strategy required a smaller engineering footprint, as opposed to the massive project at Columbus. Multiple forts not only stretched infantry manpower, but engineering talent and labor.

Island No. 10

On March 2, 1862, Polk telegraphed the War Department: "The work is done. Columbus gone." He removed all but two 32-pounders, and salvaged quartermaster and commissary supplies sufficient for eight months. When the Federals occupied the city, Brigadier General George Cullom, himself a former military engineer, noticed that the works were "of immense strength, consisting of tiers upon tiers of batteries on the river front, and a strong parapet and ditch, covered by a thick abatis, on the land side." The focus now went to Island No. 10. If it was to be held, then New Madrid, on the Missouri side of the river and north of the island, also had to be occupied. Although Beauregard never visited the town, his engineer instincts led him to make suggestions. He envisioned a triangular fort defense, with two of the works anchored on the river and one in advance. The gorges (entrances) should be palisaded, so that the makeshift Confederate gunboats from New Orleans could fire into them should they be captured. Brigadier General John P. McCown, commanding the brigade-size garrison, arrived on February 26, 1862. He immediately expressed disgust at what he found. "I find that little or nothing has been done by Captain Gray." On the Tennessee shore, the redan, he believed, would probably not hold up. As for the island, work had only commenced the previous day. The glacial pace of the construction left the fortifications, and the rivers they purported to defend, vulnerable to the invading Federal navy.[34]

Gray did receive a new engineer—twenty-two-year-old Lieutenant Arthur B. De Saulles of New Orleans, who received his civil engineering degree from the Polytechnic School of Geological Studies in Troy, New York, and then the Imperial School of Mines in Paris. Among his many studies, he was proficient in the topographical plotting of maps. After working on the fortifications at Columbus, De Saulles was transferred to Island No. 10, where he fell wounded during the siege and recuperated in Corinth, Mississippi. After the war, he moved to Pennsylvania. In 1917, while visiting his son Jack on Long Island, New York, he witnessed his son's murder by his daughter-in-law. Already in bad health, De Saulles passed away shortly thereafter, his friends claiming that he "died of a broken heart."[35]

McCown was in a panic. On February 28, 1862, a boat brought up more laborers from Fort Pillow, but they had no tools. If the river rose much more, Battery No. 1 (the redan) would be flooded. The one ammunition magazine, apparently in an effort to give better protection, had been poorly situated below water level. Four days later the enemy was within three miles of New Madrid. Even if his 4,000-man force held against John Pope's 20,000-man army, he feared that the gunboats would close their ports and make a run past Island No. 10. "My position is critical in the extreme," he wrote. New Madrid was subsequently invested and, following a fifteen-day-siege, fell on March 14. The rebel garrison escaped, but the bungled affair fell just short of a stampede.[36]

Most of the Columbus guns, ammunition, and supplies were sent to Island No. 10. Brigadier General James Trudeau, accompanying the army advance, believed that more work was accomplished in twelve days with a thousand men than had been done with Polk's entire corps at Columbus in three months. He was most appreciative of Captain Wintters and his sapper company, who "rendered valuable assistance and executed the work now under consideration of mounting guns with great dispatch." Between February 24 and March 22, 1862, Wintter's men, with the exception of four days, either mounted guns, built platforms, or constructed gun carriages every day. While expressing his gratitude to Gray, Trudeau nonetheless added a criticism. The batteries on the high ground, back from the river, had raised gun platforms, which doubled the labor. "The platforms for river batteries where the ground is high, ought always be sunk," he wrote. "By this mode the same object is attained with less labor and the battery is far better protected." In defense, the river could, and often did, widely overflow its banks, which would have left the sunken batteries submerged. Indeed, by spring the gun platforms in the redan were three to nine inches below high water, and shortly thereafter it had to be abandoned. Thus the dilemma for engineers: rising water submerged fortifications or at the least eroded their walls, thus forcing batteries farther and farther from the river and losing the advantage of effective fire.[37]

Beauregard dispatched David Harris to report the progress of Island No. 10. The engineer conceded that neither on the island nor on the Tennessee shore were there any commanding positions, although he calculated that both were ten to fifteen feet higher than the west bank (Missouri shore). The rising gauge of the Mississippi River (Gray called it "unprecedented") was once again causing bank erosion, forcing two of the shore batteries to be moved farther back. The labor shortage, a perennial problem, had to be addressed. There were only 200 Blacks, 128 Memphis Irish laborers, and Wintter's company of forty men, which was totally occupied in mounting guns and preparing platforms. At least 1,000 Blacks were needed, with 300 axes, 500 wheelbarrows, and 200 shovels.[38]

On February 19, Federal engineers explored the possibility of cutting a canal through several bayous on the west bank, thus permitting the vulnerable troop transports to bypass the island. Through remarkable ingenuity and skill they succeeded, and the transports emerged from the swamps. The bluecoats now had the capacity to make an amphibious landing at several points below the island. Without the protection of the ironclads, however, such a maneuver remained dangerous. The heavy and lumbering vessels could only come through the main river channel. The joint operation was coming to a head. "If Island No. Ten falls, we will have to run again," wrote a rebel, "and there is no force between us and Memphis to prevent them from going to that place. The people in Memphis are badly frightened—well they may be, for only six Regiments stand between them and the enemy."[39]

Captain Scheibert was harsh in his evaluation of McCown and his dismissive attitude of the Federal canal. In defense of Gray he wrote: "He—only he—acted. To secure Confederate positions he put batteries near New Madrid to bombard the [canal] exit." While this is true, Gray never expressed alarm about the canal. An article in the *Memphis Appeal* assured its readers: "The trees in the Mississippi bottoms . . . are very large and grow close together, and sent their roots deep into the soil. This growth of our swamps and bayous presents an impenetrable barrier to any undertaking as that spoken of."[40]

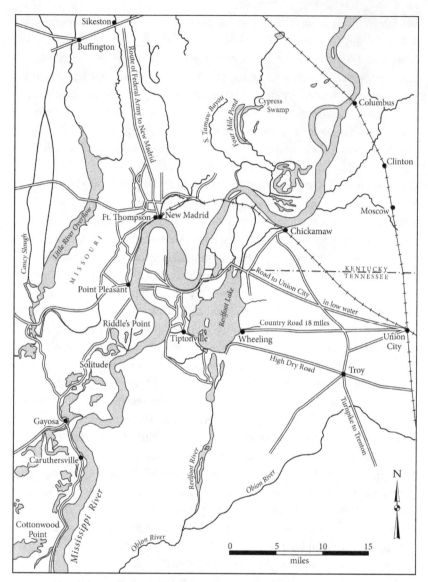

Map 1. New Madrid topographical map. The Missouri shore was drawn from existing land office surveys. The Tennessee side was created from surveys, a topographical party reconnaissance, and from the "best authorities at hand." *War of the Rebellion*, vol. 8, 146

In early March, thirty-five-year-old, Berlin-born engineer Captain Victor Von Sheliha (Viktor Ernst Karl Rudolph von Scheliha in German) arrived at Island No. 10. He had received formal training in Prussian military schools in both artillery and engineering. Jeremy Gilmer, chief engineer for Department No. 2 (to be discussed momentarily), always viewed the German with suspicion. His opinion did not change when Von Sheliha later transferred to Mobile and married a handicapped woman from a wealthy family. "Foreigners in our army are, I fear, mere soldiers of fortune," he later wrote. It did not take long for Von Sheliha to get into a row with Pillow. He told the general that he had completed his assignment, to which the Tennessean abruptly replied: "It is not so, sir. The work you report done is not done." Gilmer immediately wrote to Albert Sidney Johnston stating that if Pillow's charges were true, then the captain should be held responsible, but if not true, then "he is entitled to protection from insult." He was subsequently sent to perform a reconnaissance in northern Alabama, but later returned to Island No. 10.[41] Sometimes such reconnaissance missions involved just a handful of men, other times an entire company.

The Federal navy attacked in earnest on March 15–17. Later studying the reports, Captain Scheibert concluded that the parapets, which he over-estimated to be about twenty-four feet, were "thick but on a soft basis and wave-drenched at the escarpment." They "gave way under bombardment. Thirteen-inch projectiles exploded in the breastworks. Through the holes [breaches] came the deluge. Defenders waded in water and mud." The captain was clearly writing of the action at Battery No. 1, the redan, which took a tremendous pounding. Water came to within a half-inch of the gun platforms. Several hundred Blacks were employed in making repairs during the night.[42]

In late March, Brigadier General William Mackall assumed command of Island No. 10, "that infernal trap," as he referred to it. Von Sheliha made a report on April 3. On the Tennessee shore, Battery No. 1 (the "redan") was under water and unserviceable. The powder magazines were ample for the ammunition on hand, but required strengthening on top. The island had no magazines or bombproofs, and the men were entirely unprotected. In postwar years Von Sheliha did not hold back on his criticism. "These batteries could not be considered samples of engineering skill," he candidly wrote. "They consisted of a parapet five feet high and twenty-two feet thick; they had no traverses against the splinters [fragments] of bursted projectiles; and no service magazines . . . and being protected against the weather by tarpaulins spread over them. The main magazine for all these batteries was hardly proof against heavy shells; the laboratory had been established in an old shanty situated a little to the rear, but still within reach of the enemy's fire." On the dark and stormy nights of

April 1 and 6, two ironclads made it through the chute between the island and the Tennessee shore, sustaining little damage. Von Sheliha disgustedly wrote that thirty-eight of the fifty-one guns dismounted after opening fire, "in consequence of the badly constructed platforms having settled in the new ground." The island, with its garrison of 4,400, was captured, Von Sheliha among them.[43]

Fort Pillow

Fort Pillow was next in line. Since Beauregard never ventured to the fortification, it appears that he received his information through David Harris; he was not pleased at what he heard. About three miles east of the fort, Cole and Hatchie Creeks came to within 1.5 miles of each other. "Yet the engineers who planned the works . . . had not availed themselves to this natural advantage, and, strangely enough, instead of erecting the land defenses at the point mentioned, had placed them nearer the fort, lengthening their line more than three miles, and necessitating a garrison of nearly ten thousand men. A similar error . . . had been committed at Columbus," wrote Alfred Roman, Beauregard's ghostwriter.[44]

As matters continued to deteriorate at Island No. 10, Major General Braxton Bragg, recently arrived at Corinth, Mississippi, expressed alarm. "Unless something is done speedily for the defense of Fort Pillow I fear we shall lose the Mississippi—of more importance to us than all the country together." Brigadier General Alexander P. Stewart, commanding at the fort in late March 1862, outlined his concerns. One of the river batteries, with eight smoothbore 32-pounders, had four inches of water on the platform and three feet of water in the rear of the platform. Beauregard's smaller land defensive line "would require some days for construction." An interior line was nonetheless begun for 5,000 men, but was never fully completed. When David Harris arrived on April 1, only two batteries on the land side (Nos. 1 and 5) were operational, and they were a mile and a half apart. An additional sapper and miner company, commanded by Jules V. Gillimard, arrived from New Orleans, but by late April 1862 Wintter's and Gillimard's companies counted only forty-six present for duty.[45]

Brigadier General John Villepique's report of April 6 expressed serious and uncorrectable engineering concerns. First, the topography was such that a small force could simply not hold the position against a large one. An extensive line had already been prepared, ranging from one-half mile to one mile from the river, but a part of it was "so badly located that it would be untenable. Several hills, entirely protected from fire, command it in reverse." The line was so extensive that a garrison of 15,000 and a large number of guns would be required.[46]

Fig. 1.2. Fort Pillow. The batteries at the base and atop Fort Pillow were drawn by US Navy rear admiral Andrew Walke. Johnson, *Battle and Leaders of Civil War History*, vol. 1, 449.

On April 16, Gray, now at Fort Pillow, was killed in a boiler explosion aboard a gunboat. A year after the war, his remains were reinterred in Elmwood Cemetery in Memphis. David Harris was ordered to Vicksburg four days later. Engineer responsibilities now fell to Major Meriwether. He and the captain of the C.S.S. *Polk* took a small boat up the Hatchie River, ten miles below Fort Pillow. Before the Hatchie emptied into the Mississippi River, it came to within 1.5 miles of Coal Creek, just above the fort. The team was convinced that if a gunboat ascended the Hatchie, it could command nearly half of the distance to Coal Creek through an open field. Meriwether was concerned that if an ironclad passed the Fort Pillow batteries, it could drop anchor in a bend in the river three miles downriver and enfilade the fort without interruption. The major noted that the site engineers (apparently a reference to Lynch and Harris) had been warned about this but had taken no action. Fearing that the enemy could approach within three miles of the fort unmolested, the outer entrenched line was abandoned, except that portion between Coal Creek and the Ripley Road.[47]

The Albert Sidney Johnston-Beauregard team failed to destroy Grant's advancing army at the Battle of Shiloh on April 6–7, 1862, ultimately leading to the loss of Corinth. The rebels abandoned Fort Pillow on the night of June 5, blowing the magazines and casemates and lighting up the night sky. Fort Wright was subsequently abandoned and Memphis left to surrender after a brief naval engagement. Although criticism could, and was, leveled at the work of Confederate engineers, the loss of the Upper Mississippi River defensive line was ultimately not a failure of engineering but of strategy. As long as the river fortifications were flanked, they could be taken in reverse and inevitably fall like dominoes. The forts should have been seen for what they were—mere stopgaps to buy time. Indeed, it could be argued that the Island No. 10 fortress, constructed by a cadre of civil engineers and architects, previously experienced only in flood control, railroad construction, and surveying, for weeks held back a 20,000-man Federal army, thus buying time for Johnston to potentially win a victory at Shiloh. Due to the physical geography of the Heartland, Confederate engineering was thus an underappreciated component of the overall strategic plan. If not the lynchpin, the engineers were at least an important piece of the overall puzzle.[48]

2

SCANDAL AT THE TWIN RIVERS

FROM DOVER, TENNESSEE, north into Kentucky and extending to the Ohio River, the Cumberland and Tennessee Rivers parallel one another from three to twelve miles. Technically Leonidas Polk represented Confederate authority in Tennessee during the summer of 1861, but his jurisdiction was limited to west and north of the Tennessee River. Besides, he was pre-occupied with invading Missouri and protecting the Mississippi River. Responsibility for defending the so-called Twin Rivers thus temporarily fell to the state of Tennessee. Historians have criticized Governor Isham Harris for his dilatory response, focusing his attention on the Mississippi River, but such was not the case. In April 1861, nearly a month and a half before Tennessee seceded, state authorities turned their attention to these potential invasion routes. Loss of the Cumberland River meant the loss of Nashville, but the loss of the Tennessee River meant an open avenue all the way to northern Alabama, where the shoals obstructed further advance. Two of the river's six bridges, those at Clarksville, Tennessee, and Florence, Alabama, could be destroyed. Additionally, control of the Tennessee River meant that Federal authorities could ascend upriver and take either Columbus or Nashville in reverse.[1]

Thirty-four-year-old Adna Anderson, despite his New York birth and Union sentiments, responded to Harris's appeal for assistance in defense of his adopted state. A well-respected Nashville civil engineer with vast railroad experience, he currently served as the receiver of the Edgefield & Kentucky Railroad. Assisting him would be twenty-seven-year-old Lieutenant Wilbur F. Foster, a diminutive, northern-born civil engineer who had received notoriety for overseeing the construction of Nashville's first railroad bridge. Foster's map of Nashville proved exceptional. He could be blunt and temperamental and, while he held peer respect, he had few close friends. Foster reported to Anderson on May 10, 1861, near Dover, Tennessee, a village of twenty to thirty houses, a brick courthouse, a tavern, and a church. The team began surveying the west bank of the Cumberland River. A site was selected eleven miles south of the Kentucky state line, along a ridge north of town, with the main battery atop a 100-foot-high bluff, "just a happy gift of nature," a Union officer later remarked. Construction began with a workforce from the Hillman & Brother Iron Works.

The survey party then proceeded to the Tennessee River, where the terrain from Standing Rock Creek north was examined for several miles. "In all these surveys great care was taken to ascertain true high-water mark and note the conditions which would exist in time of floods," wrote Foster. Two sites were selected—one midway between the Paris Road and Standing Rock Creek, nearly opposite the mouth of the Big Sandy River, and the other a mile downriver at Coleman's Landing. His work complete, Anderson returned to Nashville. When the city subsequently fell on February 25, 1862, he offered his services to the Federals and ultimately became head of the Construction Corps of the Army of the Potomac. Anderson settled in the North after the war. Showing signs of insanity in later life, he committed suicide in 1889.[2]

Desiring military corroboration, the governor dispatched a West Pointer, weathered old Brigadier General Daniel Donelson, to examine the positions. He had no practical engineering credentials and had been out of the academy for thirty-six years, raising the question posed by one historian: "[H]ow much engineering knowledge these former cadets retained from their days at the USAMA?" Donelson nonetheless possessed political gravitas—speaker of the Tennessee House of Representatives and nephew (by marriage) to President Andrew Jackson. He found the Cumberland River fort, which ultimately bore his name, the "strongest position on the Cumberland near the State line." One of the state engineers, thirty-one-year-old Captain (state rank) John Haydon, former civil engineer for the city of Nashville, expressed his disgust. The location, he asserted, was based on the "ill advised idea that the low water of the river for the greater part of the year was guarantee enough [to prevent gunboat passage]."[3]

Donelson next examined the Tennessee River position; he was not pleased. There were no good sites in Tennessee, and he expressed a desire to go into Kentucky. Although no location was specified, he perhaps had in mind the so-called "Birmingham narrows," where the two rivers closed to within three miles of one another. Such a move would have advanced the Confederate line

Fig. 2.1. Bushrod Johnson. Heading the Corps of Engineers of the Tennessee Provisional Army, Johnson made the final decision on where to locate both Fort Henry and Fort Donelson. Courtesy Alabama Department of Archives and History, Q50.

twenty miles into Kentucky. It proved moot, however, for Harris, at least at that time, strictly honored that state's neutrality. Donelson thus chose Kirkman's Landing, five miles north of Anderson's sites.[4]

Another potential site was Pine Bluff, on the east bank of the Tennessee River three miles from the state line and opposite Pine Bluff Landing, on the Kentucky side of the river. An engineer described the bluff as a "very steep, rocky, and large hill." Even if heavy guns could be drawn to the summit, which would be problematic, the barrels could not be sufficiently depressed for effective firing if the Union ironclads hugged the shoreline.[5]

The final decision came on June 9, when Colonel Bushrod Johnson, a West Pointer heading the state Corps of Engineers, arrived on the scene. Quiet and retiring and, according to an officer who knew him at the time, not overly impressive, Johnson had previously served as superintendent of the Western Military Institute at Nashville, where he also taught mathematics and engineering. Interestingly, the subject of engineering had not been his strong suit while at the academy. He adopted Donelson's selection of Kirkman's Landing, between Panther and Piney Creeks. He rejected Anderson's site because of the high ground in the rear of Coleman's Landing that would command the position and that would require a large force to defend it. He considered Kirkman's "much superior to it in a military view." Kirkman's Landing had no commanding heights, and his decision was made "exclusively on account of the rear defenses." Focused only on the land side, he made no mention of the flood plain. Johnson left the final decision to Harris but warned, "Too much delay has already occurred." The die had been cast; Kirkman's was approved. The subsequent weak and poorly sited earthen work became Fort Henry, named for Clarksville, Tennessee, politician Gustavus A. Henry.[6]

Fort Henry was not without some advantages. It offered a direct line of fire for over a mile and, according to William Preston Johnston, the creeks and freshets offered protection on the land side. There were two direct twelve-mile roads—the Telegraph Road and the Peytonia Furnace Road, connecting Henry on the east bank with Donelson on the Cumberland west bank, meaning that the garrisons could be mutually supportive. Of course, any flooding that kept the Federal infantry at bay on the land side would have the same effect for Confederate reinforcements.[7]

Lieutenant Foster was the first to raise engineering concerns. Detailed to survey the area, he recognized that the fort would be below the high-water mark and therefore subject to probable flooding. He reported this to the governor, who curtly informed him that he was too young to criticize the work of older engineers. Disgusted with the response, the prickly lieutenant requested reassignment to his regiment—the 1st Tennessee, which Harris granted. Foster subsequently accepted a commission in the Confederate engineer corps and was sent to East Tennessee. Precisely why Colonel Johnson failed to see the obvious has long been questioned. Perhaps he figured that virtually all places on the east bank were in a flood zone. Nonetheless, as the river rose not only would the fort be flooded, but also the gunboats would be raised to a higher level, placing them above the fort parapet.[8]

Jesse Taylor, a former naval officer, also noticed the "extraordinarily bad judgment" in the fort's location. Even the most casual observer could see the watermarks on the trees, indicating that "we had a more dangerous force to contend with than the Federals, namely, the river itself." He reported this to the state authorities, who advised him that the decision had been made by competent engineers. Realizing his ideas "conflicted with those entertained by a 'West Pointer,'" Taylor acquiesced. In subsequent conversations with local citizens, however, he discovered that even in an ordinary February the fort would be under two feet of water. He again notified state authorities, who brusquely informed him that the fort was now under Confederate jurisdiction, and that his concerns should be forwarded to Polk. Taylor sent several

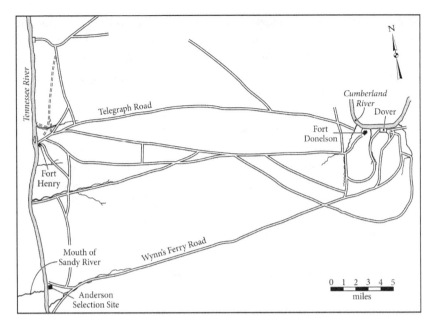

Map 2. Forts Henry and Donelson.

warnings, but the bishop-general only passed them on to department headquarters at Bowling Green. Frustrated with the run-around, Taylor realized that it was "now too late."[9]

By August 1861, three other state engineers, in addition to Foster and Haydon, had been assigned to the Fort Henry project. Lieutenant Felix R. R. Smith, twenty-two, grandson of General James Robertson, founder of Nashville, was an assistant engineer for the city of Nashville. Thomas L. Estill, fifty-five, of Winchester was the son of a prominent surgeon and "a scholar in his profession (civil engineer)." Colonel Johnson wrote of fifty-five-year-old John G. Mann of Wilson County: "I am favorably impressed with him." The construction of the fort was primarily done by troops of the 10th Tennessee, a "Regiment well used to the construction of canals and Railway," wrote Haydon. The work, begun on July 1, was mostly completed by the end of August.[10]

With Haydon, Mann, and Smith in Middle Tennessee, and Pickett, Rucker, and Lynch in West Tennessee, the embryonic engineer corps of the Tennessee Provisional Army was left with only one engineer—thirty-six-year-old Major Achilles Bowen, who was assigned to East Tennessee. A West Point graduate who had excelled in his engineering classes, Bowen had refused an army commission and became a civil engineer and farmer in Columbia, Tennessee. Upon receiving his Confederate commission, he briefly performed engineering work in East Tennessee before being assigned to staff work in Nashville. He soon resigned, however, "feeling that too low an office had been given one of his training and well-known military ability." Another West Pointer, Charles C. Rogers of Pulaski, was offered a commission, but he refused and never reported for duty. This left Johnson with six engineers in the state army.[11]

Haydon again expressed his dismay. "Upon no point upon the River [Tennessee] could a less favorable place have been chosen," he wrote. "The terra plain was subject to overflow, the highest point being fully eighteen inches [below water] and the lower parts seven and a half feet." A twelve-foot-deep slough, which paralleled the river on the fort's eastern face, frequently flooded. On three occasions the exasperated Haydon determined to go public with "loud complaint and forcible denunciation." Fearing that the memorandum would fall into the hands of the enemy, however, he "tore up the papers each time." Despite having earlier professed to Johnson "to have made a study of military engineering," Haydon began to question himself—"Where was my military knowledge?"[12]

Red flags continued to be raised. Colonel Adolphus Heiman, commanding the 10th Tennessee, expressed his apprehensions, and he backed them up with numbers. "At high-water stage of the river the water backs up into Panther Creek on the north and Lost Creek on the south 2½ miles, and at this stage the lower part of the fort is not free from overflow, being 7 feet 6 inches lower than the highest point." The west bank (the Kentucky side of the river) commanded the fort at a level of 170 feet. A 3,000-foot-long ridge to the northeast had an elevation of sixty feet above the parapet and could become a base of operations for the enemy. Colonel Johnson nonetheless expressed no concern and, on September 15, wrote that Fort Henry was "a good enclosed work, with bastion fronts."[13]

In desperate need of engineers, Johnston transferred Lieutenant Joseph Dixon from Polk's headquarters to the Twin Rivers. Polk loudly protested, and three orders had to be issued before he finally released the lieutenant. Dixon, a thirty-two-year-old from Athens, Tennessee, graduated from West Point in 1858, third in his class. He served in the topographical engineers in the Departments of Oregon and Washington. Although he was the engineer in charge at the Twin Rivers, he became the de facto site manager for Fort Donelson. By the time he arrived at Fort Henry on September 17, the work was nearly completed. Although he reported that the fort was "not at the most favorable position," he did not believe it should be abandoned, but rather strengthened. He then proceeded to Fort Donelson. Some 600 slaves arrived from farms in Kentucky and Tennessee, and work slowly progressed. A friend of Dixon's remembered that he kept an exhausting schedule, working at times during the night. He always spoke of work "just completed," or to be "laid off for tomorrow."[14]

Gilmer

In mid-October 1861, Major Jeremy Gilmer arrived in the West to become chief engineer for Department No. 2. A graduate of West Point, his profile included assistant instructor of engineering at his alma mater, Mexican War veteran, and engineering stints in New York City, Savannah, Georgia, New Mexico, and San Francisco, California, where he was stationed when the war broke out. With eight years of practical experience in the field in map making, surveying, and fort construction, few surpassed him in either army. The major and his wife, Louisa, had been close personal friends to Henry Halleck and his wife, but the relationship forever soured because of the war. Unfortunately, Gilmer was not pleased with his assignment. He desired to be in Savannah, Georgia, where Louisa lived and the weather, or so it was believed, was more suitable to his rheumatism. Gilmer's orders were to oversee the Twin Rivers, Clarksville, and Nashville, entrusting the details to subordinates. He specifically was to "Arrange a plan of defensive works for Nashville, and urge them forward by all means you can command."[15]

Gilmer arrived in Memphis on October 14, where he received orders to proceed to Johnston's headquarters. He optimistically, perhaps naively, wrote his wife that he thought it "possible we will winter in Louisville—the place you love so much." He arrived at the department headquarters at Bowling Green on the 15th to discover the true state of affairs and to form an opinion of Johnston: "The General is not very communicative." He spent the next few days battling a severe head cold, but he eventually warmed up to Johnston, who "shows a disposition to be very friendly with me." By the twenty-third, Gilmer was surveying the region south of Bowling Green with the assistance of two engineers, one of whom was Lieutenant Smith, of the Tennessee state engineers. "Some of these [works] are far advanced toward completion. Others will have to be built to make our position as strong as it ought to be," he concluded.[16]

Gilmer began a river tour on October 25, starting at Clarksville, Tennessee, forty-one miles below Fort Donelson. The strategic importance of the town lay primarily in the railroad trestle/drawbridge of the Memphis, Clarksville, & Louisville Railroad, which was completed only a few weeks before Gilmer's arrival. The railroad connected to the Louisville & Nashville Railroad just south of Bowling Green. If the Clarksville bridge was destroyed, communications

Fig. 2.2. Jeremy Gilmer. Serving as chief engineer of Department No. 2 from October 1861 to April 1862, Gilmer held a mixed view of General Albert Sidney Johnston. Courtesy The American Civil War Museum, Richmond, Virginia.

between Johnston's center and left wing would be severed. Paying close attention to the flood plain, the chief engineer suggested a battery at the mouth of the Red River. Also, to guard against an enemy incursion west of Bowling Green, in an attempt to turn Polk's right, he recommended a line of trenches on the high ground north of Clarksville that commanded the Hopkinsville Turnpike.[17]

He then turned the project over to thirty-four-year-old, Irish-born Captain Edward B. Sayers, who had served as assistant engineer of the state of Missouri prior to the war. Active in the Missouri State Guard, he had drawn the layout for Camp Jackson. In 1861 Sayers got into a row with Brigadier General Daniel Marsh, challenging him to a duel. The engineer fired a shot and missed and Marsh shot in the air, thus ending the episode. Assisting Sayers at Clarksville would be Captain Walter J. Morris, described as "an intelligent civil engineer." Ultimately Forts Clark (Red River), Levin, and Terry (railroad bridge), all at Clarksville, were constructed.[18]

Around the 31st, Gilmer arrived at Dover. Time was clearly now of the essence, but the chief engineer gave Fort Donelson little more than a perfunctory inspection. He spent a day at Fort Henry, where he was impressed with the work. His only recommendation was one long recognized by others—fortify the commanding west bank of the river. On January 1, in company with Dixon and in the midst of a drenching rain, he took a boat north on the Cumberland River into Kentucky, where a project was underway at Line Island and Line Port under the supervision of civil engineer T. J. Glenn to sink barges with stones to obstruct the channel. Influenced by an old Nashville steamboat captain, Dixon put great stock in the undertaking. Gilmer, having absolutely no knowledge of the river, gave his nod of approval. Colonel Heiman thought the project to be a complete waste of time "in a river which rises from low-water mark at least fifty-seven feet, and which I myself have often known to rise at least ten feet in 24 hours." Dixon nonetheless maintained high confidence. Gilmer preferred the Line Port site, fifteen miles below Donelson, and apparently considered relocating the fort. With time of the essence, however, it was decided to simply strengthen the existing fortification; he had little choice.[19]

Gilmer's subsequent November 3 report to Johnston was astonishingly lacking in detail. Indeed, Fort Henry, clearly the weakest link, was not even mentioned. In another communication, he related to Senator Henry that he thought Fort Henry to be "in fine condition for defense, the work admirably done." The politician saw what the engineers either failed to see or refused to report, that Fort Donelson was "in very bad condition." "The guns at Donelson are wholly unprotected," he informed Johnston on November 7. "The abatis is finished at Donelson, but no work done to protect the guns."[20]

The parapet of the main river battery at Fort Donelson was only eleven feet wide, totally inadequate by the standards of the day. Although there were many variables, the available artillery penetration tables made it clear that parapets should be at least one-third to one-half wider than the anticipated penetration of a shell. Redoubt No. 4 at Columbus, which was in the direct line of fire, was eighteen feet thick. The main battery at Donelson was not in a direct line of fire, but at an eleven-foot thickness with a one-third reserve would allow for a penetration of little more than seven feet. Even a relatively small 18-pounder firing at 880 yards could penetrate nearly four and a half feet. The Federal ironclads would, of course, be using far large guns, such as 8-inch Dahlgrens firing shells weighing nearly fifty-three pounds. The thin Confederate earthen ramparts would later prove insufficient. The parapet at Fort Henry, at least according to a Cairo correspondent who later measured it, was only eight to ten feet thick. None of the Tennessee state engineers had studied under Mahan, but certainly Gilmer and Dixon had done so, and these deficiencies should not have gone unnoticed.[21]

In postwar years, Beauregard expressed criticism. Gilmer should have recognized the weakness of Fort Henry and relocated it, presumably to Kentucky. The conclusion was ill-founded. To relocate the fort at that point would have lost five months of work. Additionally, if Fort Henry were moved, Fort Donelson would also have to be rebuilt, *if* the forts were to be mutually supportive. William Preston Johnston disclaimed criticism of the engineers, but then proceeded to criticize them. "By bringing the combat to the water level, it deprived the fort [Henry] of any advantage of elevation," he wrote. Of course, if he (or his father) had bothered to actually go to the site, he would have seen that the closest elevated area was fully three-fourths of a mile from the river, well out of the 1,500- to 2,000-yard range of the heavy artillerists. For better or for worse, the battle with the ironclads would be fought on the river's edge at Fort Henry. This meant that the Union ironclads could (and subsequently did) close within a direct fire range of 500 yards, delivering devastating salvos with their huge Dahlgren guns.[22]

Between November 4 and December 19, Gilmer was absent from the Twin Rivers. "I went down the river [Cumberland] for the purpose of seeing the character of the points selected for defense and to suggest such additional works as in my judgment might be necessary," he wrote his wife. Even in his private letters he raised no alarmist language. The chief engineer divided his time between Clarksville and Nashville, drawing Dixon away from Fort Donelson for several days and, to Haydon's utter disgust, removing him from Fort Henry for a month to make surveys around the Tennessee capital. "For two days past," Gilmer wrote his wife from Nashville on November 13, "I have been down along the river [Cumberland] below here, riding and walking through about the roughest country I have seen. Tomorrow I expect to go again." The next day, still fighting his recurring rheumatism, he made a grueling thirty-mile trek around the city. The exhausting schedule continued throughout a cold and wet mid-December.[23]

Gilmer began making inquiries to the chief quartermaster as to pay vouchers. He notified railroad executive V. K. Stevenson on November 26 that he had "employed a number of civil engineers as assistants" for the work around Nashville. One of these (Sayers) had been sent to Clarksville and had to be paid out of Gilmer's own private funds, for which he needed reimbursement. He notified Johnston that payments should be made monthly, not bi-monthly.[24]

Historian Thomas Connelly concluded that Gilmer was "happy to be in Nashville, and went about his duties at a leisurely pace." He also criticized Gilmer's defensive line north of Nashville, rather than along the steep bluffs of the south bank. Gilmer did this ostensibly to protect Edgefield. "The cream of Nashville society lived at Edgefield, and the young, impressionable Gilmer was undoubtedly pressured to locate the defensive line north of the village."

To locate the defenses on the south bank, however, would have placed virtually all of Nashville and its vital bridges within artillery range. On the face of it, it would indeed appear that Gilmer became obsessed with the Nashville defenses. It must be remembered, however, that General Don Carlos Buell's 70,000-man Army of the Ohio could have turned Johnston's right at Bowling Green by way of Glasgow, Kentucky. Even Connelly begrudgingly admitted that Johnston had to have an auxiliary line. Did Gilmer have to personally oversee the Nashville project? No, but the department was so huge that simply having him dart from one place to another would have been a waste. More to the point, in Gilmer's visit to the Twin Rivers he gave no indication that he would have made any significant changes even had he devoted his full attention to the project.[25]

The issue of labor proved an ongoing problem that was never totally resolved. Gilmer candidly thought that detailed soldiers were preferable to Black labor. Unfortunately, only 200 soldiers were fit for duty at Fort Donelson as of November 21. "I do not think that the labor of troops and slaves can be combined to any advantage," he concluded. Appeals were made to plantations, but with scant response. Gilmer informed Johnston that all slaves had been hired out till year's end. On December 6, G. O. Watts, the assistant engineer in charge at Nashville, wrote that he had only seven Blacks at work. It was estimated that 5,000 slaves were required—1,500 at Nashville, 1,500 at Clarksville, and 1,000 each at Forts Henry and Donelson, but there were never more than 200, 300, and 600 respectively.[26]

Gilmer also had to spar with interfering local commanders. Throughout November, Dixon was overwhelmed with work at Fort Donelson—mounting guns, lining off a maze of trenches, felling trees for abatis, and seeking labor. He was, therefore, taken aback when Pillow, in Columbus, Kentucky, ordered him to proceed to Fort Henry and make surveys of the Tennessee River. This order may unwittingly have been caused by Johnston. Only days earlier he had ordered Polk to tend to the west bank of the Tennessee opposite Fort Henry. "Lieutenant Dixon, who is familiar with the country, will be able to point out the proper position. No time should be lost," he wrote. Gilmer complained and had the order reversed. Haydon, who at the time was in Nashville drawing the specs for the drawbridge for Fort Henry, was sent instead. A week later, Brigadier General Lloyd Tilghman, commanding at the Twin Rivers, ordered T. J. Glenn to suspend operations on obstructing the Cumberland River at the downriver shoals. "It will be impossible for me to rely upon any work being done properly if each subordinate brigadier-general be allowed to suspend operations ordered by me," Gilmer wrote Johnston. He clearly had Johnston's ear; the project was continued.[27]

Connelly continued his litany of criticism. Gilmer was "unhappy with his rank," "showed no interest in his work on Johnston's line," was "homesick and unhappy with his assignment," held "an indifferent, unpleasant attitude," "did not think there was any real need in hurrying," "considered himself an expert" in strategy, "wasted much time in sulking" over Johnston's lack of an offensive, and "behaved like a frustrated infantry commander." Connelly's critique was unnecessarily harsh. Venting to his wife aside, Gilmer labored tirelessly, often suffering from severe rheumatism in his body, hand, and neck. The engineer, continued Connelly, also exhibited "inexcusable neglect of his duties at Forts Henry and Donelson" by getting absorbed in Dixon's river obstruction project. The historian, of course, had the vantage of hindsight, knowing that the river project failed and that the Federal attack ultimately came at the Twin Rivers. Lacking a crystal ball, Gilmer had to prepare for *all* possibilities. Even had Grant's attack on the forts failed, Buell could still have marched on Clarksville or Nashville.[28]

In early December the chief engineer was in Nashville searching for large-scale training camps for six to eight Tennessee regiments. To Dixon's request that he come to Fort Henry to help him locate the earthen work on the commanding west bank, Gilmer answered: "[I]t is out of the question for me to go to the Tennessee [River] now." Interestingly, it was the normally inept Pillow who looked into his crystal ball and saw what Gilmer either failed to see or for which he expressed no alarm. The Federals would "use their large water power to capture Fort Henry," and then destroy the Tennessee River railroad bridge, thus dividing Polk's and Hardee's Corps.[29]

Gilmer, in Bowling Green by December 20, continued his demanding schedule. The weather began to improve, drying the roads and increasing the chances of a Federal offensive. He began examining the fords of the Barron River. The rains returned the first week in January 1862, however, turning the roads into ribbons of mud—"winter campaigning in Kentucky is next to any impossibility," he wrote his wife. On the 5th, he returned from a reconnaissance along the Barron River—"It snowed and I was almost frozen." He spent the next day riding with Johnston and Brigadier General Simon B. Buckner in search of additional defensive positions around Bowling Green. "I cannot tell you our strength but our numbers are small. If we are defeated [at Bowling Green] Nashville is lost," he confided.[30]

When Beauregard arrived at Bowling Green, he philosophically disagreed with Gilmer's extensive earthworks. All that was needed, he argued, was a *tete de pont* on the north bank of the Barron River and a small fort to protect the bridge on the south bank. While the projection may have been an underestimation, his concept was correct. Bowling Green could be easily flanked. There was no reason for Buell to attempt a direct assault against extensive works. He was well aware of the fortifications; indeed, a map showing all of the forts and the number of guns in each was printed on the front page of the *New York Herald*. As at Columbus and Fort Pillow, the Louisiana general once again preferred a mobile to a static defense. Extensive earthworks were nonetheless undertaken. To cover the Glasgow and Scottsville Turnpikes to the Barron River, Gilmer requested 800 picks, 1,500 shovels, 1,500 axes, and 500 wheelbarrows.[31]

Back at Fort Henry, the long-anticipated arrival of 300 slaves from North Alabama neared. They were to work on the west bank fortification that would bear the name Fort Heiman. As a part of the advance team, Alabama politician James E. Saunders arrived on January 17. He expressed shock that no work had commenced. He was told by Haydon that Tilghman had "not passed on the plan." The labor finally arrived and, while they had plenty of axes, they lacked a sufficient number of picks and shovels. Of the fifty wheelbarrows, only ten remained serviceable by the end of the first day. Nonetheless, the works were far advanced, though incomplete, by the end of the month. It was eventually decided to send the slaves home, fearing they would panic in the event of an attack.[32]

When word got back to the department headquarters about the true state of affairs at Fort Heiman, the normally reserved Johnston openly expressed his ire. His frustration was aimed at Polk. He turned to William D. Pickett, then at Bowling Green, and as he paced exclaimed: "It is most extraordinary. I ordered General Polk four months ago to at once construct these works, and now, with the enemy on us, nothing of importance has been done. It is most extraordinary, most extraordinary." The engineer recalled that he "never saw him so wrought up before."[33]

On January 20, 1862, Gilmer returned to Clarksville. The latest intelligence indicated that Buell would thrust from the Green River to Clarksville, interposing himself between Hardee and Polk. The state of Tennessee issued a desperate, though fruitless, appeal for 500 more slaves. Word also floated that the Federals planned an amphibious operation, with the infantry disembarking fifteen miles below Fort Henry. They would then march toward Murray, Kentucky, turning Polk's right and threatening his rear. Gilmer believed the roads to be too muddy for such a maneuver, but he suspicioned that Henry W. Halleck, commanding Pope's and Grant's armies, would "have to make a showing." On the 23rd, word arrived of the Confederate defeat at Mill Springs in eastern Kentucky, but Gilmer remained confident of the Bowling Green defenses—"I feel good about them."[34]

With indications of an attack, Gilmer returned to Fort Henry on January 31. For the next two and a half days, in company with Tilghman, he examined the four-acre fort and thirty-five acres of rifle trenches. The two rode to Fort Donelson on February 3. Several changes were made at that time, including raising the embrasure sides (cheeks) with sandbags. This proved a poor substitute for reveting embrasures with mounds of packed clay and sand, something taught by Mahan at West Point, and a flaw which should have been noticed by Gilmer the previous November. Packing, of course, took time and labor, something which engineers and work details often could not afford. Sandbags were subject to being torn with incoming rounds, but could quickly be replaced. For whatever reason it had been missed, and now Gilmer was involved in a feeble patch job. At noon the next day unexpected firing could be heard back at Fort Henry. Learning from a courier that a large enemy force had disembarked at Bailey's Landing, Gilmer and Tilghman returned to the Tennessee River fort, arriving at 11:30 p.m. A heavy rain that night flooded the low roads, filled the river with driftwood, and inundated the lower part of the fort with waist-deep water. William Preston Johnston referred to the flooding as "unprecedented"; the more appropriate word would have been predictable. The nightmare portended by Haydon and so many others had come to pass.[35]

About 11:00 a.m. on the 6th, the gunboats attacked. The naval guns of the enemy made deep penetrations into the parapet. There were no breaches, but the sandbag cheeks were utterly smashed. So swollen was the river that the gunboats actually fired in a downward trajectory at the fort. "Major Gilmer, with glass in hand, quietly watching the effects of each shot gave an occasional direction," noted a correspondent for the *Memphis Appeal*. In less than an hour it was all over and Tilghman lowered the flag. The Federals floated a lifeboat through the sally port to receive the surrender. Indeed, Jesse Taylor later concluded that if the attack had been made two days later, the magazines would have been under water and the fort would have surrendered with barely a shot. The garrison, along with Gilmer, escaped to Fort Donelson, while a 100-man rearguard remained and was captured, Haydon and Joseph A. Miller among them. A northern correspondent could not resist writing that the fort had been improved by "the scientific skill of engineers taught at West Point, at the expense of the Government of the United States." Many southerners would have added that it was West Pointers, not state and civil engineers, who devised the original "harebrained scheme" to locate the fort in a flood zone.[36]

Tilghman could not refrain from denouncing the location of Fort Henry. "The history of military engineering records no parallel to this case," he groused. "Points within a few miles of it, possessing great advantages and few disadvantages, were totally neglected and a location fixed upon without one redeeming feature." He never specified the superior location "within a few miles." Tilghman also lamented that the fort could be enfiladed at three points along the shore and three or four points on the land side. In postwar years, William Preston Johnston caustically responded to Tilghman. He noted that the brigadier, himself a civil engineer, never once voiced objection as to the fort's location until after its surrender, even though he was widely known for his outspoken opinions.[37]

The surrender of Fort Henry placed Johnston in a dangerous situation, something which Beauregard clearly understood. Grant could take his army in transports up the Tennessee River and approach Nashville from the rear (a sector which lay undefended), while Buell's army could threaten Hardee's Corps from the front. With the fleet protecting the Union line of communications, Polk would be helpless to offer support. In that scenario Fort Donelson could be entirely bypassed, although Grant's flank would be exposed as he advanced inland. Grant, however, had other plans.[38]

Johnston became convinced that Fort Donelson could not hold against the ironclads. He initially made plans to abandon Bowling Green, but then reversed his decision. He instead reinforced the Donelson garrison with an additional 12,000 troops. Even his normally sympathetic biographer declared the decision to be a "monumental mistake." Johnston exacerbated the problem by placing two inept political generals—Pillow and John Floyd, in command. "They were popular with the volunteers," he weakly explained. Johnston was never completely clear why he decided to reinforce the garrison. If he believed the fort could not stand against the ironclads, then why send more infantry? Historian Benjamin Cooling, and other historians, concluded that the reinforcements were meant to be a delaying force, to buy time for Hardee's Corps to withdraw from Bowling Green. Unfortunately, "he never made that crystal clear to any of his brigadiers."[39]

With such a large number of reinforcements arriving, the original defense line of Fort Donelson, between Hickman and Indian Creeks, had to be expanded. Unfortunately, tall trees had already been felled to create an abatis, and this would now complicate movement with the expanded line. The new line, between Hickman and Lick Creeks, stretched for over two miles. The troops frantically worked day and night, although the expanded line was never fully completed. Indeed, Buckner complained that his division line was not one-third complete. What works did exist amounted to a few logs rolled together and slightly covered with dirt. A Texan described his trench as being two feet deep and with dirt thrown over saplings to make a five-foot protection. The main river battery had no traverses, and the lower river battery had no bombproof.[40]

Privately, Captain Dixon (his promotion had finally been approved) began to express his fears. He whispered that unless the garrison was strongly reinforced, the position could not be held. He was also overheard to say that if the garrison surrendered, he would move to Central or South America. Throughout Wednesday, February 12, Dixon, now commanding the river batteries, drilled the gun crews. Early the next morning, as the artillerymen continued practicing, an ominous sight appeared downriver—the U.S.S. *Carondelet* rounded the bend. An hour-and-forty-minute duel ensued, but little damage was inflicted on either side. The ironclad's next to last shot, however, struck the check of a 32-pounder, disabling the gun and blasting the crew, Dixon among them. An iron screw tap struck the captain in the left side of his face and head, killing him instantly. His body was taken to McMinn County, Tennessee, for burial.[41]

The main naval attack by five gunboats occurred on February 14. The timberclads in rear of the ironclads blasted "curveted shot, which passed over our works, exploding in the air just above," reported Gilmer. One of the ironclad shots penetrated three log cabins in the fort, equaling nearly thirty inches of timber. Once again, using sandbags for the embrasure cheeks, rather than mounds of packed clay, proved costly. Lieutenant Colonel Milton A. Haynes, an artillery officer, reported: "The heavy shot tore away the cheeks of our embrasures, throwing the sandbags upon the banquette [interior steps], and exposing our gunners to the direct shot of the enemy." To the astonishment of many, however, the Confederate gunners defeated the ironclads. Meanwhile, Grant's army, 27,000 strong, invested the fort from the land side. Confederate indecision squandered an opportunity for the 18,000 troops to break out. Although 2,000 or so, including Gilmer, escaped, the bulk of the garrison surrendered. The loss proved stupefying and Johnston would never fully recover.[42]

A Chicago correspondent later toured the captured fort and could not resist making comments on the engineering. "There was a manifest attempt on the part of the engineer who planned it to see what could be done in angles," he observed. "I counted more than thirty. It was more an exhibition of fancy on the part of the engineer than a display of common sense or scientific attainments." He thought the fort "the most irregular thing imaginable. Its like was never before constructed." In terms of the land defenses, "it needed but a glance to see that there had been defective engineering. With the force they had there was too much ground to look after. A more skillful engineer would have selected commanding points on the ridge and thus concentrated strength."[43]

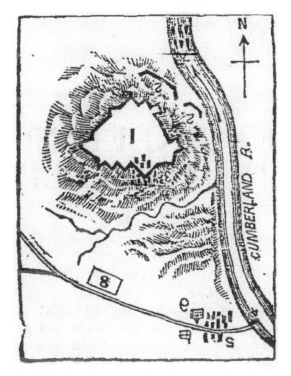

Fig. 2.3. Design of Fort Donelson. The odd configuration of Fort Donelson so captivated a northern correspondent that he took measurements and drew a rough sketch for his readers. *New York Tribune,* February 19, 1862.

The most egregious engineering decisions at the Twin Rivers—the location of Fort Henry in a flood plain, the over-reliance on obstructing the Cumberland River, the failure of raising the parapet with sandbags rather than packed clay, and the overly extensive land defenses—were all decisions made by West Pointers. Did officials err by not relocating the forts into Kentucky once the neutrality was broken? Not necessarily. The Federals launched a small joint army-navy operation in the Eddyville area on October 31, 1861. Any attempt at relocating the forts would have been disrupted at that time. In the end, faulty engineering became another nail in the coffin that helped seal the fate of the Confederate Heartland. The river highways were now open to Union invasion.

3

ENGINEERING IN THE FIELD

WITH THE LOSS of the Upper Mississippi Valley and the Tennessee and Cumberland Rivers, the strategic focus shifted from rivers to rails. As early as February 1862, General Mansfield Lovell, commanding at New Orleans, suggested a troop concentration at Corinth, Mississippi, a vital north-south, east-west rail junction. During the spring of 1862 through the end of the year, the role of the engineer corps thus transitioned from static defensive positions to mobile operations. Engineers were now called upon for reconnaissance, map sketches, bridge demolition, and repair of roads, fords, and small bridges. Although the army had two companies of sappers and miners, the role of Confederate engineer troops had not fully come of age.[1]

Transition

With the fall of Fort Henry and a similar fate anticipated for Fort Donelson, Johnston, on February 7, 1862, issued orders to quietly begin the evacuation of Bowling Green. It would be the 16th, a full nine days later, before the rain-soaked, sickness-ridden column straggled the sixty miles into the Tennessee capital. Therein lay the danger of Bowling Green. If the probing attack of the Federal navy on February 14 had actually succeeded in passing the Donelson batteries, the ironclads could have been in Nashville before the Confederates. It was a moot issue as matters developed. The gunboats were beaten back and, even after the surrender of Fort Donelson, Halleck ordered the ironclads to pause at Clarksville so that Buell's army could take Nashville. It was the only good bit of luck that Johnston had during an otherwise disastrous chain of events.[2]

The shortage of engineers continued to bedevil the theater commander. Haydon, Joseph A. Miller, and Walter J. Morris were all captured at the Twin Rivers, and Dixon was killed. Captain Pickett transferred from Columbus to Bowling Green, where he joined Hardee's staff as assistant inspector general, thus losing another designated engineer. On the Mississippi River defensive line, Lieutenant Snowden had been accidentally killed at Columbus, De Russy transferred to Louisiana, De Saulles wounded, Von Sheliha captured at Island No. 10, and Gray killed at Fort Pillow. Rowley, Wilbur Foster, Mann, Smith, Estill, and Charles Foster had all been reassigned to East Tennessee and Harris to Vicksburg, Mississippi. Bushrod Johnson had taken a field command. Following the fall of Fort Pillow and the surrender of Memphis, Montgomery Lynch became a refugee without a command. Either permanently or temporarily, Polk's and Hardee's Corps had been deprived of nineteen engineers.[3]

For the time being, only a small cadre of engineers remained in Middle and West Tennessee—eight to be specific. Gilmer had escaped capture twice, and Meriwether and Sayers managed to get out of Fort Pillow just before it was abandoned. Lieutenant Henry N. Pharr, twenty-eight, an Arkansas civil engineer, initially served in the 1st Arkansas Mounted Rifles until wounded in the Battle of Wilson's Creek, Missouri. Upon his recovery, he was transferred to the engineers and had been working at various positions along the Mississippi River. Thirty-six-year-old Lieutenant George B. Pickett, brother of William and also a civil engineer, was present at Bowling Green. Twenty-four-year-old native-Kentuckian George M. Helm reported for duty in Nashville. Some years earlier he had worked under Minor Meriwether as a civil engineer on the Mississippi & Tennessee Railroad. By the 1850s Helm had relocated to Washington County, Mississippi, where he was involved in drainage and flood protection. Gilmer assigned him to Gower's Island, twenty miles below Nashville, to reduce the Cumberland River channel to seventy feet. In postwar years, Von Sheliha insisted that even after the fall of Fort Donelson "a few deep and heavy blasts" of the high riverbank would have filled the narrow channel with masses of solid stone, completely obstructing the Cumberland River. Given the failure of Dixon's obstruction project, such a conclusion seems unlikely. James D. Thomas, twenty-seven, a former Kentucky civil engineer, also served on Johnston's staff at Bowling Green.[4]

Arguably the most qualified and certainly the most unique engineer at Bowling Green was thirty-seven-year-old Captain James Nocquet (pronounced NO-KAY), formerly of the French army, having served in Algiers. He had lived in the United States a number of years prior to the war, although he still spoke broken English. Nocquet began his American career as a staff architect for the Illinois Central Railroad and later went into private practice. Gilmer would write of the captain: "[F]ew men present greater qualifications."[5]

All was in commotion at the Bowling Green depot, noted a correspondent, as sick soldiers and terrified civilians crowded aboard the last departing trains. As Johnston withdrew on February 13, he ordered the engineers to blow the Barron River bridges. Lieutenant George Pickett, with twenty-five kegs of blasting powder, oversaw the workmen who detonated both the railroad and pike bridges. As the Federals approached, a *New York Herald* reporter noted: "The splendid iron railroad bridge and turnpike bridge have been blown up and burned." Leaving Nashville to its fate, Johnston ordered the engineers to destroy the wire suspension bridge and the wooden railroad trestle. The column withdrew to Murfreesboro, Tennessee, where George Crittenden's defeated division united from East Tennessee. What happened next has long been debated. Beauregard insisted that he coaxed Johnston to bring his troops to the rail junction of Corinth, where the two wings of the army could unite. In postwar years, Governor Harris wrote that, while at Murfreesboro, Johnston related his intent to concentrate at Corinth. While the issue of Corinth was clearly discussed, subsequent events reveal that a definite decision had not yet been made.[6]

Writing from Murfreesboro on February 26, Gilmer apologized to his wife for his dilatory letter writing. The railroad and pike bridges at that place had washed away, and he had been occupied in rebuilding them. Gilmer frankly did not know where Johnston was headed—to Decatur, Alabama, Chattanooga, Tennessee, or to points south. The next day he confided that half of the army would go to Shelbyville and half to Chattanooga. Much has been made of Johnston's February 25 dispatch to Secretary of War Judah Benjamin stating that he intended to move Hardee's Corps "to the left bank of the Tennessee . . . in order to enable me to co-operate or unite with General Beauregard for the defense of Memphis and the Mississippi." The salient words, of course, were *co-operate or unite*—there was a big difference. It seems inconceivable that the chief engineer would be intentionally left in the dark if a definite decision had been made. Johnston ultimately decided to move to Corinth, where Beauregard was assembling Polk's Corps.[7]

Gilmer learned of the move to Corinth while at Shelbyville on March 3; he vehemently opposed it. Worried about Virginia, he believed that Johnston should have gone east to Chattanooga; his concerns were real. Halleck, with headquarters in St. Louis, now held overall Union command of Pope's, Grant's, and Buell's armies. He could have suspended army operations at Island No. 10, leaving it to the navy, and sent Pope's 20,000 troops to reinforce Grant's Army of the Tennessee, which had taken transports south and encamped at Pittsburg Landing, Tennessee. This would have freed Buell's army to march directly on Chattanooga, Tennessee, which lay ripe for the picking. Gilmer increasingly became disenchanted with Johnston, and for that matter the other generals whom he had encountered, noting "among them there is not a Napoleon."[8]

Gilmer arrived at Decatur, Alabama, on March 9, and it did not take him long to express his disgust—"This place is a poor apology for a town and the sooner we can get away the better." The major was also becoming irritated with his failure to receive a colonel's commission, especially since he had heard of Mackall's promotion. If he had received no news by March 12, he vowed to tender his resignation. No news came and, venting aside, he continued with his duties. While a scout mapped northern Alabama and Mississippi, between the railroad and the Tennessee River, Gilmer proceeded to examine the Mississippi River defenses. He arrived in Memphis on March 24. From there he took a steamer to Randolph and Fort Pillow. At the last he left "two or three young engineers" to strengthen the fortifications. Although not named, one was probably Lieutenant H. A. Pattison, a topographical engineer known to have assisted Gray. Following the capture of Clarksville, Sayer also arrived at Fort Pillow.[9]

Once on the south bank of the Tennessee River, Johnston faced a predicament. Halleck had ordered Buell's army to juncture with Grant's at Pittsburg Landing. Johnston had no pontoon train and, on March 18, he ordered "silent preparations" to burn the bridge at Florence, Alabama. The question remained about the bridges at Decatur, Huntsville, and Bridgeport, Alabama. If these bridges were destroyed, Johnston could only recross his army by use of ferries or by marching as far east as Chattanooga. He therefore left no order for the engineers to torch them. Once northern Alabama was occupied by the Federals, however, they surely would have to destroy the bridges if they were forced to withdraw, which is precisely what happened at all three locations. One way or the other, these bridges, once in the hands of the enemy, would be denied the rebels. Under these circumstances, Johnston should have destroyed them.[10]

Throughout March, the Army of the Mississippi, as it was styled, assembled in Corinth. Added to Polk's, Hardee's, and John C. Breckinridge's Corps (the last actually a division) was a newly assembled corps from the Gulf Coast under General Braxton Bragg. The assembled hodgepodge totaled about 44,000 troops, nearly equal to Grant's army, although the Federals also had the support of the gunboats *Tyler* and *Lexington*. The troops were amateurs, but were united in their disdain of the Yankees and in the understanding that the tide of events in the West had to be reversed.

Accompanying Bragg was his chief engineer, twenty-five-year-old Captain Samuel H. Lockett. Virginia-born but Alabama reared, Lockett attended West Point, graduating second in his class. Upon graduation, he briefly served as an assistant professor at the academy before doing engineer work for the Regular Army in the Eighth Lighthouse District at Pensacola, Florida. Lieutenant Silvanus W. Steele, a thirty-one-year-old civil engineer and surveyor from Middle Tennessee, and, according to Lockett, "a splendid scout and as brave a man as ever lived," was appointed to Bragg's staff. Thirty-three-year-old New York-born Lieutenant Jason M. Fairbanks lived in Florida at the commencement of the war and accompanied Bragg from Pensacola. A qualified civil engineer in surveys, public works, and with the Florida, Atlantic & Gulf Central Railroad, he tried in vain to get an engineer's commission. He spent the next year underemployed as one of Bragg's clerks before finally being properly placed.[11]

Realizing the dire shortage of technical personnel, Beauregard requested the transfer of two engineers from Virginia—Leon J. Fremaux and John M. Wampler. The War Department moved slowly on the request, however, and Fremaux did not arrive until just before the fall of Fort Donelson and Wampler weeks later. Born in Paris, forty-year-old Fremaux moved to America as a child. As a civil engineer, he worked for the Louisiana Board of Public Works and lived in Baton Rouge at the start of the war. His adoring daughter would later recall this time as their best days. His true passion was as a watercolorist, and he would become known after the war for his remarkable full-color prints drawn between the 1850s–1870s entitled *New Orleans Characters*.[12]

Thirty-one-year-old Maryland-born John Morris Wampler was known to his friends by his middle name. He served as a topographical engineer in the US Coast Survey, where he revealed his remarkable skills in hand-drawn maps of Galveston Bay in Texas and the northwest shore of Massachusetts Bay, the maps based upon a complicated series of triangulations and survey markers. Embittered by his lack of advancement, he resigned from the US Coast Survey in 1853. He gained civil engineering experience as the chief engineer of the Baltimore City Water Works. Wampler later returned to Loudoun County, Virginia, where he had grown up, to perform surveys for an emerging railroad company. His extraordinary map skills soon came to the attention of Beauregard, who requested his transfer to the West.[13]

Shiloh

Johnston determined to strike Grant's Army of the Tennessee before Buell's Army of the Ohio, marching from Nashville, could unite. Beauregard directed

Fremaux to draw a map showing the region between Corinth and Pittsburg Landing. Completed on April 3, 1862, the one inch per five-mile scale was not entirely accurate. The Savannah Road and Ridge Road intersected at a place called Michie's. On April 5, as the Confederate army snaked to the battlefield on different roads, a traffic jam occurred at this vital intersection. On the map, however, the "Mickey" house, as Fremaux labeled it, was shown several miles distant from the intersection. Additionally, Lick Creek did not flow entirely parallel to Owl Creek, but bowed to the east, thus creating a wider battlefield than the southerners expected. Johnston ordered additional maps of the roads around Corinth, but the deluge of rain washed out the main Corinth Road. As soon as the water receded, Fremaux wrote, he would "make the maps as ordered."[14]

Johnston sent engineer officers, such as Lockett, Steele, and Helm, on reconnaissance missions to Pittsburg Landing. During the Battle of Shiloh, April 6–7, 1862, engineers were used for scouting and essentially as orderlies. Given their scarcity, placing them in the direct line of fire performing secondary tasks seems at best curious and at worst wasteful. The army's seven engineers nonetheless accompanied the staffs of their generals, the dearth of technical talent notwithstanding—Gilmer on Johnston's staff, Major Minor Meriwether on Polk's, Lockett and Steele on Bragg's, Fremaux and Helm on Beauregard's, and Nocquet on Breckinridge's.[15]

During the dawn hours of April 6, Lockett and Steele were ordered to take a company of cavalry and make a reconnaissance on the Union left, giving updated reports to Bragg. Steele had scouted the Federal camps on several occasions and knew the country thoroughly. The officers approached on foot to within several hundred yards of the Federals. In some camps they found few signs of activity, only smoldering fires. In other camps troops lined up for roll call, while others cleaned their rifles for Sunday inspection and cooks prepared breakfast. The bluecoats had clearly been caught unaware. As the battle opened, however, and the Federals withdrew, often in panic, Lockett discovered that the Federal line overlapped that of the Confederates. A Federal "division," actually David Stuart's brigade, could attack the rebel flank. At 11 a.m., as no reinforcements arrived, Lockett reported directly to Johnston. "Well, sir, tell me briefly and quickly as possible what you have to say," the general tersely said. Based on Lockett's report, two additional brigades were sent to the right. So thickly wooded was the battlefield that Pickett recalled that he ended up as a courier for several different generals. Lockett spent the balance of the day carrying orders and making recons for Bragg. The most dramatic moment came when the general ordered Lockett to take the flag of the 4th Louisiana forward at the Hornet's Nest. The banner was snatched from Lockett by Colonel Henry W. Allen, saying: "If any man but my color-bearer carries these colors, I am the man." Johnston fell mortally wounded on the afternoon of April 6, leaving Beauregard in command.[16]

Grant was reinforced by Buell's army that night, and the Confederates grudgingly relinquished the field on the 7th, having sustained brutal casualties. Lockett spent most of the day collecting men, having his horse shot from under him in the process. He gathered about a thousand troops from a half dozen regiments, labeling the hodgepodge the "Beauregard Regiment." Late in the day Gilmer was shot in his right arm, fracturing a bone. He recuperated in Georgia, thus severing his immediate ties with the western army. As the army withdrew to Corinth, Bragg ordered a battalion of infantry to "work the roads." "I left my only engineer, Captain Lockett, with General Hardee; five [engineers], at least, could be well occupied," he reported.[17]

Back at Corinth, the Louisiana general braced as Halleck's combined army group of Grant's, Buell's, and Pope's armies, 90,000 strong, inched forward. The Army of the Mississippi counted only 35,000 troops to counter them. The Confederates took up positions, with Polk and Breckinridge on the left, Bragg in the center, and Hardee on the right. Extending Hardee's right would be Earl Van Dorn's newly arrived Army of the West from Arkansas, actually only an undersized 13,000-man corps. Lockett was ordered to find appropriate camp grounds for Van Dorn's divisions around Rienzi and Jacinto, Mississippi.[18]

The Corinth defenses were laid out by Lockett prior to Shiloh and approved by Johnston. The defenses, dubbed the "Beauregard Line," extended for three miles in a semi-circle tilted to the northeast. Following Shiloh, Beauregard, who described the defenses as rifle pits and "slight" batteries, made some adjustments. The works were located behind a small creek, but behind it lay a near impenetrable mud-swamp "as wide as Broadway," according to a New York correspondent. The bluecoats expressed grave concern for the defenses, for Beauregard had "the most skillful engineers from the old United States Army." When the town was later occupied, the correspondent expressed contempt. "The 'fortifications' hardly deserve the name. They were the simplest description of breastworks. . . . They were not half as strong as those constructed by our troops in a single night."[19]

Finding Gilmer's replacement would prove problematic. Lacking an available engineer in the Regular Army, the appointment curiously went to Virginia colonel John Pegram, a West Point graduate with no designated engineering background. In the Regular Army he had served routine frontier stints and in 1861, during the Rich Mountain Campaign, he was captured. He was now exchanged and awaiting an assignment. The position should have gone to Lockett, but Pegram had the personal endorsement of General Robert E. Lee. A prejudice for Virginians thus trumped practical engineering skill. Envisioning a withdrawal from Corinth, Beauregard dispatched his new chief engineer to Tupelo, Mississippi, to scout the ground. Seemingly out of his depth, the young colonel asked Beauregard to come down personally "as the ground is rather different from what we imagined from the map."[20]

As officers prepared their Shiloh after action reports, Fremaux began drawing a pen and ink watercolor map of the battlefield to accompany Beauregard's report. The scale, 5/8 inch per mile (1:101,376), had relief shown by hachures, a form a shading that showed terrain steepness. A slightly different battlefield map, unsigned but also approved by Beauregard, was perhaps also drawn by the Frenchman. Meanwhile, a pioneer battalion under the leadership of Major Meriwether was organized. The battalion, although officered by engineers, was comprised of detailed infantry and probably performed only manual labor. The unit was dissolved in late July 1862. Lieutenant George B. Pickett assumed command of the sole sapper and miner company, which was organized back at Bowling Green.[21]

Wampler arrived in late April and began preparing a map of the Corinth defenses, a task seriously hampered by a lack of instruments. He began by borrowing Fremaux's compass. The engineer was also directed to lay out a new defensive line on the right flank along the high ground three miles outside of town. Breckinridge detailed 500 men to dig the trenches and battery positions. Wampler and Lockett kept Beauregard apprised of the progress in nightly meetings. Unfortunately, Wampler was stricken with sickness, confining him to bed for three days. Beauregard, increasingly nervous of the Federal approach to the right, sent Wampler to line off more earthworks in that direction. Additionally, with a view of an inevitable withdrawal, Pegram, Lockett, and Wampler began examining the roads to the south. The 1st Louisiana was detached as pioneers to work on the roads, but skirmishing slowed progress.[22]

De Saulles arrived at Corinth but, not yet fully recovered from his Island No. 10 wound, he did not report for duty until mid-April. He was then directed to line out defenses between Monterey, Tennessee, and the Memphis & Charleston Railroad. Ordered to Grenada, Mississippi, on May 5, he prepared a map of northwest Mississippi and examined the condition of the Mobile & Ohio Railroad. Lieutenant James K. P. McFall of Maury County, Tennessee, a "jolly good fellow with a bald head," reported to Corinth in May. Though not a civil engineer, he had experience in construction on the Nashville & Decatur Railroad and was assigned to the engineers. Appointed to the pioneer battalion, he was ordered in July 1862 to take Company B and report to Vicksburg. Lieutenant John G. Mann, one of the original Tennessee state engineers and now in East Tennessee, was detailed to Nathan Bedford Forrest's cavalry.[23]

On May 16, Stephen Wilson Presstman (he preferred to be called by his middle name), one of Beauregard's Virginia transfers, also arrived at Corinth. The thirty-one-year-old, Delaware-born captain had served as a railroad civil engineer in Alexandria, Virginia, prior to the war. As an officer in the 17th Virginia, he had been wounded at Blackburn's Ford. Presstman became chief engineer for Joseph E. Johnston at Centerville, Virginia, in February 1862. It was in that position that he met Beauregard, who subsequently requested him in the West. On May 20, Presstman and Wampler scouted the Federal trenches near Farmington, at which time they came under fire, one shot barely missing the newly arrived Virginian. Another engineer arriving that spring was Lieutenant John W. Green. Formerly a civil engineer with the Vicksburg, Shreveport, & Texas Railroad, he had served as adjutant of the 4th Louisiana Battalion before transferring to the engineers.[24]

In late May, the Army of the Mississippi evacuated Corinth and withdrew fifty miles south to Tupelo, Mississippi. Beauregard was replaced by Bragg and strict discipline imposed. While many of the troops groused, Lockett expressed satisfaction. "These strong measures are the only ones that will do our army any good and from this time on all cases of desertion, lawlessness disobedience and unmilitary conduct is going to meet with speedy and severe punishments," he wrote his wife. "We are going to have no more playing soldier in Genl Bragg's army and I hope we will soon see a very beneficial change." Lockett, while constructing two bridges south of Tupelo over the Tombigbee River at Aberdeen, was ordered to Vicksburg. Wampler and Lieutenant Morris, recently paroled from his Fort Donelson capture, continued the Tombigbee project, with Pickett's sapper company and forty-four impressed slaves providing the labor. A shipment of tools arrived, although the ordered pile driver from Mobile, Alabama, proved unnecessary. Within eleven days a permanent bridge and a floating bridge had been completed.[25]

The Kentucky Campaign

On June 22, 1862, Bragg informed the War Department that Halleck had divided his forces, sending Buell's 30,000-man army east to Chattanooga along the line of the Memphis & Charleston Railroad. Bragg planned to get in his rear, but a lack of transportation prevented a rapid movement. He thus began transporting his 26,000 infantry by rail to Mobile and then to Atlanta and

Chattanooga, while his artillery, cavalry, and wagons trekked overland through northern Alabama. The Army of the West, reinforced with Breckinridge's division, remained in Mississippi to confront the remaining Union forces under Grant. New Orleans, Louisiana, surrendered to the US Navy on April 25, with the garrison withdrawing to Vicksburg.

The western engineer corps likewise divided. Meriwether, along with four others, remained in Mississippi. Sayers informed Gilmer: "Fremaux is left at Tupelo, in charge of the Mississippi Corps (5 officers) to accompany Genl. [Sterling] Price north." Fremaux recalled: "At Tupelo I was designated to accompany General Bragg in Kentucky, but my knowledge of the Mississippi River in Louisiana, caused my orders to be changed to the command of General Breckinridge in Baton Rouge." Pegram, once in Chattanooga, was reassigned as chief of staff to Major General Edmund Kirby Smith, commanding the Department of East Tennessee. Replacing him would be David Harris, who had returned from his Vicksburg stint. Wampler became chief engineer to Polk's right wing and Presstman to Hardee's left wing. Assisting Presstman would be Captain Sayers and Lieutenants Pharr and Green.[26]

Buell's tedious advance east along the Memphis & Charleston Railroad allowed Confederate raiders to create havoc on Union communications and supply lines. Troops of the 1st Michigan Engineers labored a month in repairing the 700-foot-long Elk Creek bridge of the Nashville & Chattanooga Railroad, which did not become operational until July 12. The next day Forrest's cavalry attacked the garrison at Murfreesboro, capturing a thousand prisoners and destroying the Stones River bridge. Eight days later Forrest struck all three bridges over Mill Creek near Nashville. There was seemingly no end to the soft targets that could be hit. On the Nashville & Decatur Railroad alone there were twelve significant bridges in a forty-five-mile stretch. Similar vulnerabilities existed along the Nashville & Chattanooga. Buell was forced to put his men and animals on half-rations. Raiders were simply overwhelming the repair capabilities of Union engineers, infantry fatigue parties, and 600–700 hired Blacks.[27]

Raids were not the only problem stalling Union operations. Buell had no pontoon train and the Tennessee River above Stevenson, Alabama, had to be crossed. The Michigan engineer troops were assigned to make a pontoon bridge for the 1,400-foot crossing. The work was expected to take a week, but a month and a half later the job was still incomplete. Some 300,000 feet of oak lumber were needed, and only one local sawmill could cut the long timber required. Oakum and pitch had to come from Louisville, and the nails sat somewhere in a Nashville depot. By mid-August, five companies of the 1st

Fig. 3.1. David B. Harris. As chief engineer of the Army of the Mississippi, Harris commanded engineer operations during the 1862 Kentucky Campaign. Courtesy Cook Collection, The Valentine Museum, Richmond, Virginia.

Michigan Engineers were repairing the railroad from Nashville to Louisville, with the other six companies remaining in northern Alabama. Buell's advance ground to a halt. On August 24, work on the pontoon bridge was called off and most of the boats burned. As the Federals withdrew to Nashville, they destroyed three bridges of the Nashville & Chattanooga Railroad, including those over the Elk and Duck Rivers.[28]

Bragg arrived in Chattanooga and, while awaiting his artillery and transportation, took the opportunity to restructure his command. The army was divided into a right wing under Polk and a left wing under Hardee. "I am connected with no particular Div. but belong to the Left Wing generally immediately under Gen'l Hardee's command," Lieutenant Green wrote his wife. Bragg also ordered a sapper and miner company for each wing. George Pickett's company served under Polk, and Captain G. W. Maxson's company in Hardee's command. Pickett's outfit comprised one engine driver, two bridge builders, one laborer, two track layers, three brick masons, twelve miners, and nineteen carpenters, with two civil engineers (Pickett and T. S. Newcomb) as officers. Harris served as army chief engineer, with Wampler, Sayers, Morris, and Thomas with Polk's wing, and Presstman, Green, Pharr, and Helm with Hardee's wing. Sayers took time on August 3 to give Gilmer, then in Richmond, an update of affairs: "We have a very nice set of men now in the Corps. Only two of the commissioned officers cannot do anything as engineers."[29]

The *Chattanooga Daily Rebel* sounded the alarm that the defenses of East Tennessee needed to be strengthened. "Has any able engineer been sent here to examine the country and report what works are needed?" While there is no evidence of engineer activity in Chattanooga at that time, the Knoxville district had a small working staff. Rush Van Leer, twenty-three, a former Nashville civil engineer, was at Cumberland Gap. The Knoxville office included John Moore, a civil engineer, and two draftsmen—Charles Foster and Conrad Meister. Captain Edmund Winston's company of sappers and miners was divided, with a detachment placing obstructions in Big Creek Gap and the balance of the company at Cumberland Gap. Winston's detachment at Big Creek Gap was subsequently attacked, losing five killed and thirteen captured, the prisoners being exchanged in September.[30]

Working directly on Kirby Smith's staff as draftsman was a fascinating aristocratic (son of the mayor) New Orleans civil engineer and architect by the name of William "Will" Freret. The twenty-eight-year-old had studied civil engineering in Britain and returned to New Orleans, where he took up architecture. He built a set of speculative homes (still there today) in the Garden District to attract wealthy buyers, only to have the war break out. He lost nearly everything in what became known as Freret's folly. He was detailed from the prestigious Louisiana Washington Artillery (5th Company) to serve on Smith's staff. He became a preeminent southern architect after the war, and many of his structures survive. Also on Smith's staff was Colonel William R. Boggs, a West Point graduate who worked for two years with the topographical engineers upon graduation before being assigned to ordnance duty. At the beginning of the war he served with Lockett on Bragg's staff at Pensacola, although he appears to have been bored with engineer work. He became embittered about being passed over by others for higher rank, and he and Bragg turned out not to be a fit. He thus resigned his Confederate commission and began working as a military engineer for the state of Georgia. The governor later sent Boggs to assist Smith, although he served only in an advisory capacity. He nonetheless represented another trained engineer.[31]

Bragg now went on the offensive. Kirby Smith's two divisions, reinforced with Patrick Cleburne's and Preston Smith's brigades from Bragg, reached Cumberland Gap on August 14. The plan was to invest the gap and then return

Fig. 3.2. Captain William "Will" Freret (standing, *far right*). Freret was photographed with a group of engineer officers serving on Kirby Smith's staff in the Trans-Mississippi. Rodenbough, *Photographic History of the Civil War,* vol. 1, 105.

and cooperate with Bragg in a move on Buell's army. The remaining portion of Winston's company of sappers, led by thirty-one-year-old Lieutenant George R. Margraves, an East Tennessee civil engineer, accompanied Cleburne's brigade. Smith instead left a division to watch the Union force in the gap and proceeded to Lexington, Kentucky. Lacking unified command, Bragg could only warn Kirby Smith to not advance too far north before he could be supported. Kirby Smith, on August 30, scored a smashing victory at Richmond, Kentucky, capturing a 5,000-man Federal division on the field of battle.[32]

Engineers in the Van

The engineers spearheaded the army into Middle Tennessee, with the engineer officers, the two companies of sappers and miners, and the 1st Louisiana, the last providing cover from bushwhackers and serving as pioneers. The main army moved out of Chattanooga on August 28. Once at Sparta, Bragg received the encouraging news of Kirby Smith's victory at Richmond. Bragg had planned a move into Kentucky via Harrodsburg, but a report of severe water shortages led him to shift toward Glasgow. He divided his army at Sparta, with Polk's wing moving toward Gainesboro, Tennessee, to cross the Cumberland River, and Hardee's wing veering northwest toward Nashville. If Hardee could hold Buell's attention, then Polk, by hard marching, could reach Glasgow and sever Union communications with Louisville. Wampler arrived at Gainesboro at midnight on September 6. The Cumberland River, north of town, narrowed at that place and usually could be forded. At daylight Wampler explored the two fords below the town. The crossing nine miles above proved out of the question. "The path I have chosen is practicable except for 2 miles which will require much work," he reported. Polk's column would have to march two miles over a mountain and then up the river another four miles. Harris provided similar information to Hardee's wing, which marched via Carthage.[33]

The two corps of the Army of the Mississippi united at Glasgow in mid-September, but Smith was still 200 miles to the northeast. Bragg then received news that Buell was at Bowling Green, thirty miles to the east and too close for comfort. For some days Harris and Sayers had been scouting far in advance of the army, some fifty-five miles northeast to Bardstown by way of New Haven, deep in enemy territory. They reported the pike to be in good shape, although the bridge over the Barron River, twenty-two miles from Glasgow, had been burned. The ford remained in good shape, but the banks on both sides were steep and would require repairing for wagons and artillery. Literally every source of water—ponds, creeks, springs—in the parched region from Bardstown to Louisville via Shepherdsville was plotted. Similar scouting was also done for the creeks and ponds on the Louisville Road via Mt. Washington. On the 9th, David Harris garnered valuable intelligence from citizens concerning road conditions and mileage. Rather than uniting with Smith for a planned move on Cincinnati, Bragg, on September 15, suggested a combined attack on Louisville. The plan derailed, however, when Bragg decided instead to march north to capture the 4,000-man Federal garrison at Munfordville on September 17.[34]

Engineer officers remained scattered on September 23—Wampler at Bardstown preparing seven maps for Bragg and Polk, while Pharr and Morris, with their escorts, rode east toward Danville. Pharr reported that some of the roads were good, but others were uneven and would require work. The only good source of water was Chaplin Creek, which ran through Perryville. On the 29th, Wampler rode twenty miles from Mount Washington on the Bardstown Pike toward Louisville. Matters almost went terribly wrong on October 1, when he rode 250 yards beyond the picket line and encountered a large force of Federal cavalry; he narrowly escaped toward Mount Washington. October 2 found Wampler at Bardstown, where he briefed Harris on his reconnaissance notes.[35]

Following the subsequent setback at the Battle of Perryville on October 8, in which Pharr was captured, and the failure of Kentuckians to rally in support, Bragg withdrew to Tennessee. "I will leave it to the newspapers of criticizing Gen'l Bragg's career," Lieutenant Green informed his wife following the campaign, but he still knew "of no General of our service, under whom I would prefer fighting." The troops arrived in Knoxville in late October in a near starving condition. Army strength had dwindled to a dangerously low 26,156, with some regiments mustering barely a hundred men. "Our armies here [Knoxville] are gradually, but certainly, melting away, whilst we are getting no re-enforcements. No recruits, and cannot see a source from which they are to come," Bragg reported. Davis supported the army commander, but the command structure began to show cracks. Meanwhile, Bragg used the respite to reorganize his army, which he christened the Army of Tennessee. "I hope that I will be more proud to claim to be one of this new Army than of the old A. M. [Army of the Mississippi]," an officer admitted.[36]

While at Knoxville, Norquet, now a major, superseded Harris. The order had actually been cut on October 7, but it seems unlikely that Harris received the news until his return to Tennessee. Beauregard requested Morris at Charleston, South Carolina, but there was work to be done before he departed. Nocquet had been serving with Breckinridge's division in Louisiana. The division was now on its way to Murfreesboro, but Gilmer in Richmond desired the Frenchman to hurry ahead "with the least possible delay." "The local knowledge of Capt. N in the vicinity of Louisville and other important points in the state is so great I have assigned him to duty at Head Quarters," Gilmer notified Breckinridge, but the campaign ended before his arrival. Lieutenant Green expressed his disgust. "I don't expect to remain in this army long," he wrote his wife on November 8. "Capt. now Major Harris our late Chief Eng'r has been ordered to Genl. Beauregard and has been succeeded by Capt. [Major] Norquet—a fool of a Frenchman that I have no patience with." He immediately applied (unsuccessfully) for a transfer to Beauregard's South Carolina department.[37]

Stones River

Bragg now shifted his army via Chattanooga and Tullahoma to Murfreesboro, thirty-five miles southeast of Nashville. Breckinridge's division, returning to the parent army after its defeat at Baton Rouge, arrived first. Lieutenant Rowley, having served from one end of Tennessee to the other, was transferred from the Knoxville engineer's office to Murfreesboro. Also arriving was Lieutenant Henry C. Force, a thirty-year-old Alabamian who had previously been assigned to the Army of Northern Virginia. Force attended Columbian College (later Washington University) and in 1850 served on the US Boundary Commission. From 1853 to 1855 he attended Harvard University and later became a civil engineer on an Alabama railroad connecting Tuscaloosa and Selma. Described as "well-bred and a thorough gentleman," his brother Manning, also a Harvard graduate, was a brigadier general in the Union army. He would unfortunately die of consumption in 1874 at the age of forty-two. Upon his arrival in the West, Force was immediately put to work mapping Murfreesboro and its environs. Sayers, on November 10, was sent to Decatur, Alabama, to inspect the Decatur and Bear Creek bridges of the Memphis & Charleston Railroad and arrange a contract for their rebuilding.[38]

Nocquet scouted seven and a half miles north along the Lebanon Pike, between Murfreesboro and Stones River. There was some concern about this sector, as an 8,000-man Yankee division was reportedly at Bird's Mill, less than a day's march away. The Frenchman reported the countryside as rolling, level, and virtually all cleared, and he saw no advantageous military positions. The best place to meet the enemy, if it came to that, would be north of the river. "The banks of this stream are rocky, with bluffs about 25 feet high, which makes the passage a little difficult for an army to pass; besides the north bank near the pike commands all the country south," he concluded.[39]

By December, Nocquet had been reassigned as chief engineer of Department No. 2, with headquarters in Chattanooga. It has been suggested that Bragg did this as a demotion, disliking the Frenchman's broken English. Such was not the case; Gilmer, now heading the Engineer Bureau in Richmond, made the appointment, not Bragg. The army commander preferred a return of Harris to head the Chattanooga office, but Gilmer denied the request. "Major Nocquet has had more experience as a military engineer than anyone serving in the Western Department and is intelligent in his profession. There is no available officer of sufficient experience to replace him," Gilmer wrote. Harris remained in Charleston, where he died of yellow fever two years later.[40]

Norquet would be taking with him two engineers—Lieutenant Thomas and twenty-eight-year-old Lieutenant John C. Wrenshall. Once in Chattanooga, the major submitted color drawings and specifications for a Russian-style pontoon with a wooden frame (four by eighteen feet, with a two-foot depth) covered with canvas. He further proposed 100 pontoons to a company, as compared to sixteen in the French model. Gilmer balked at the experimental design, considering it the wrong time "to make a grand experiment apparently at variance with past experience."[41]

For the time being, Wampler was tapped as the acting army chief engineer. He promptly assigned officers on reconnaissance missions, furnishing Bragg with map tracings of the vicinity. Unfortunately, a chronic bowel infection and a bout of rheumatism caused him to take a thirty-day furlough at the end of December. He would be succeeded by Captain Silvanus Steele as acting chief. Two additional engineers arrived and were assigned to the army staff. Born in Boonesboro, Arkansas, forty-four-year-old Andrew H. Buchanan taught civil engineering at his alma mater of Cumberland University, just up the road in Lebanon, Tennessee. Lieutenant James McFall also arrived in Murfreesboro, having returned from his assignment in Vicksburg.[42]

Meanwhile, Major General William S. Rosecrans, commanding the Union's reorganized Fourteenth Army Corps (to be later renamed Army of the Cumberland), petitioned for additional engineer officers. He assigned the 1st Michigan Engineers exclusively to railroad work, and organized a 1,600-man pioneer brigade, to be comprised of half laborers and half mechanics. Despite the pioneer label, the brigade performed more than manual labor. The chief engineer, James St. Clair Morton, who was not afraid of challenging standard engineering dogma, trained his men in bridge building, mapping, construction, pontoon bridges, and road repair, thus essentially serving as engineer troops. Wooden pontoon boats (bateaux) were fabricated in Cincinnati, and one company trained as a pontoon company. The end result was an organization months ahead of their Confederate counterparts.[43]

President Davis arrived in Murfreesboro on December 12; the visit was not social. Vicksburg was being threatened and Davis, believing Virginia and Mississippi to be the prime Federal targets, decided to transfer Major General Carter Stevenson's 8,776-man division from Bragg's army as reinforcement. Rosecrans soon got wind of the departure. This news, coupled with the raid of John Hunt Morgan, with exaggerated estimates of 6,000–12,000 men, resulted in Rosecrans taking advantage of a weakened Confederate army to launch a winter offensive. The 1st Michigan Engineers had 110 men in the regiment in prison for mutiny over a pay dispute at that time, but the balance of the troops were busily engaged in repairing the Mill Creek bridge outside Nashville.[44]

In the Battle of Stones River (December 31, 1862–January 2, 1863), Bragg initially deployed his army on either side of Stones River. Having been in Murfreesboro six weeks prior to the battle, he surely knew about (or should have known about) the commanding high ground, specifically McFadden's

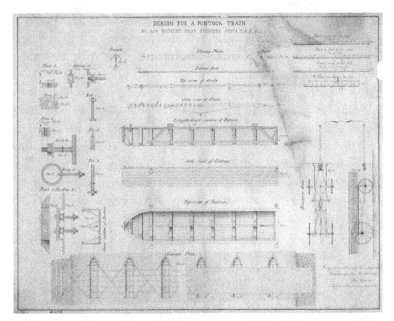

Fig. 3.3. Design for a pontoon train. James Norquet submitted his pontoon train plan to Jeremy Gilmer in December 1862. Gilmer subsequently rejected it as too experimental in a time of war. The stain is on the original copy. Courtesy Jeremy Gilmer Papers, Southern Historical Collection.

Hill on the west bank. Had Bragg occupied the hill he would have controlled the approach along the Nashville Pike. The hill was subsequently crowned by massed Federal artillery. Bragg's early morning attack on the Union right on December 31, 1862, routed Alexander McCook's corps. The Confederate juggernaut ultimately ground to a halt, however, as it encountered stiff Union resistance along the railroad embankment on the Union right and the Round Forrest on the left. On January 2, 1863, it was discovered that the Federals had crossed Stones River at McFadden's Ford and occupied a high ridge on the east bank. In an attempt to regain the position, Breckinridge's division was shredded by Union batteries on McFadden's Hill. Having lost over 10,000 casualties in three days of fighting, Bragg had no choice but to surrender the battlefield. He initially abandoned the entire Duck River line, destroying the bridge at Wartrace, and withdrew to Tullahoma, but he later returned Polk's Corps to Shelbyville.[45]

As the army withdrew, Bragg sent an engineer to investigate the fords at Elk River, apparently to cross some of his wagons. Silvanus Steele submitted his report on January 5. There were two good fords, one three feet deep at

Estill Springs and another (two and a half feet) at the Vaughn residence about a hundred feet downriver. The latter "will need causewaying, to prevent miring the wagons." Bragg ordered Polk to dispatch his sapper and miner company to repair Vaughn's Ford.[46]

The Army of Tennessee now went into winter quarters. The engineers had provided vital assistance throughout active campaigning, yet challenges remained. As their counterparts in the Army of the Cumberland expanded and became more sophisticated in their organization and operations, Bragg's engineer department was still in its infancy. The engineers would have to transition to their more expansive role, and time was not on their side. It also remained to be seen if Bragg could get beyond traditional concepts to see his engineers as partners in strategy rather than mere orderlies and scouts.

4

CONFRONTING CHALLENGES

NO MAJOR BATTLES were fought in Tennessee during the first six months of 1863, both sides being content to let the issue of Vicksburg be resolved before making a move. Bragg's army went into winter quarters, followed that spring by a period of training and recruitment. Despite the lengthy respite, the engineer corps remained saddled with problems of turnover, supplies, and over-taxed personnel. These challenges permeated all parts of the Heartland. The demand for maps, bridge construction, earthen defenses, and road repair stressed the struggling bureau's ability to cope.

The East Tennessee Connection

East Tennessee represented a vital part of the Heartland. The region was the strategic right flank of the Army of Tennessee, and the single rail line connecting Richmond to the West ran directly through the region. A largely underappreciated aspect of this mountain sector was the salt works in southwestern Virginia at Saltville. It represented the sole source of the product for the huge commissary depot in Atlanta, which supplied salt pork and beef for both Lee's and Bragg's armies. Similarly, the Ducktown Copper Mine in Polk County, Tennessee, east of Chattanooga on the Tennessee-Georgia state line, supplied the copper needed in the production of over 500 bronze 12-pounder Napoleon guns cast in the Confederacy, each one requiring 1,000 pounds of copper. In terms of strategy, communications, and raw materials, the sector had to be held.[1]

The region had long been a haunt for unionists. Edmund Kirby Smith, commanding the department, estimated that only six of thirty-two counties were

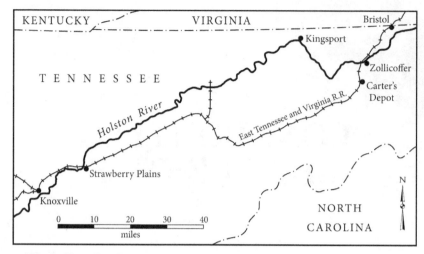

Map 3. Upper East Tennessee.

reliably Confederate. The tenuous rail line thus provided a tempting target for sabotage. Even though there was a single track, there were two different companies—the East Tennessee & Georgia from Chattanooga to Knoxville and the East Tennessee & Virginia, from Knoxville to Bristol. It was the latter that proved the most vulnerable. Pressed in Middle Tennessee, Bragg did not want to send troops to a sideshow. Smith thus counted only 7,300 troops present for duty, a fourth of whom guarded Cumberland Gap. The insurgents took advantage and burned bridges. Some were caught and hung, but the practice continued.[2]

Brigadier General S. P. Carter led a 1,500-man Union cavalry expedition in a Christmas raid in 1862 that created serious damage. The blue horsemen destroyed the 1,600-foot Holston River bridge at Strawberry Plains, fifteen miles northeast of Knoxville, the 600-foot-long bridge at Zollicoffer, Tennessee, and the 300-foot Watauga River bridge at Carter's Depot. At several places the track was pulled up and destroyed. The bridges at both Zollicoffer and Carter's Depot had been guarded by the 62nd North Carolina, but 150 were surprised and captured at the former and, after a brief skirmish, 138 surrendered at the latter. The expedition came to within eight miles of Bristol. The embarrassing affair underscored the vulnerability of the railroad.[3]

The bridges would have to be rebuilt, but it would not be quick, and the over-stretched engineer corps would once again be called upon. Seddon notified Kirby Smith on January 4, 1863, that "No time must be lost in restoring the communication." The next day Gilmer wired Captain Lemuel. P. Grant,

Fig. 4.1. The Holston River Bridge. Located at Strawberry Plains, Tennessee, the bridge was destroyed by Federal raiders in June 1863. It was subsequently repaired by Confederate engineers and the Anthony L. Maxwell & Son Company of Knoxville. This photograph was taken in 1864 while under Federal occupation. A portion of the pier survives today. Courtesy Library of Congress, LC-DIG-cwpb-02139.

the engineer in charge of the Atlanta office, to proceed to East Tennessee and inspect the damage. He arrived on January 9 and discovered that Smith had already contracted with Anthony L. Maxwell & Son of Knoxville to repair the bridges. Grant agreed that "He [Maxwell] has an organized force of skilled labor and can accomplish the work in a shorter period of time than I could possibly complete without his force and services."[4]

Gilmer promised to send Grant an assistant engineer, who turned out to be Captain John Haydon, then in Richmond. Grant, Haydon, and Maxwell met on January 11 to discuss the project. Haydon was sent to an iron furnace eight miles from Jonesboro, Tennessee, to oversee the production of pig iron. He would also superintend the bridge work, while Grant returned to Atlanta. Temporary bridges went up rather quickly, but a permanent replacement, especially at Strawberry Plains, would take time. Inclement weather halted the projects, and the timber, sawed by Wardly & Pikes of Millon, Georgia, far southeast of Atlanta, was slowed by a lack of freight cars on the Georgia Railroad. The collapse of a trestle bridge near Greenville, Tennessee, caused an eight-day delay. Haydon encountered "annoyances and indignities" in obtaining the pig iron. The end result was that two months and twelve days would be required for the Strawberry Plains bridge to be operational. A Virginia civil engineer, Captain John M. Robinson, also in Richmond, was sent to oversee the strengthening of bridge fortifications. Captain Wilbur Foster, chief engineer of the Department of East Tennessee, inspected the five major bridge crossings and several creek crossings in April. The defenses comprised blockhouses, stockades, and earthworks. It proved sufficient for small raids, but a major Federal invasion would be launched in late summer.[5]

The engineers in Knoxville during the summer of 1863 included Foster, John G. Mann, Richard C. McCalla, Felix R. R. Smith, Rush Van Leer, James S. Morrison, James Allen, A. W. Johnson, Napoleon B. Winchester, Arthur S. Barnes, Waightsill A. Ramsey, James T. Williams, Matthew Maury, and draftsman Charles Foster. McCalla was a former railroad civil engineer in East Tennessee and North Carolina. Winchester, twenty, who practiced law in Gallatin, Tennessee, joined the 2nd Tennessee and was subsequently wounded and captured at Shiloh. Upon his exchange he was assigned to the engineers. Following the war, he practiced law in Memphis, where he died at age twenty-seven from consumption. Thirty-year-old Ramsey was appointed in June 1863. Despite the fact that he was an East Tennessee civil engineer for the railroad prior to the war, he attached himself to a cavalry company from Chattanooga. He was subsequently ordered to Corinth on detached engineer duty in 1862, with the promise that he would receive an engineer's commission; it did not happen. He instead went to Fort Pillow, where he did surveys and performed reconnaissance missions for six weeks. After an extended sick leave, he was assigned to Major General Carter L. Stevenson to prepare maps at Cumberland Gap. Thirty-four-year-old Matthew Fontaine Maury Sr., not to be confused with the famed naval officer of the same name, was a Middle Tennessee surveyor.[6]

Middle Tennessee

Following the Battle of Stones River, Bragg established a new defensive line, with Hardee's Corps at Tullahoma and Polk's at Shelbyville, Tennessee. Morale increased as the ranks were filled with conscripts. "Well cousin, our country

Table 4.1. East Tennessee Railroad Defenses

PLACE	RIVER	RAILROAD	DEFENSES FOR
Charleston, Tenn.	Hiwassee	East Tenn. & Ga.	75 men
Loudon, Tenn.	Tennessee	East Tenn. & Ga.	200 men
Strawberry Plains, Tenn.	Holston	East Tenn. & Va.	300 men
Carter's Depot, Tenn.	Watauga	East Tenn. & Va.	200 men
Zollicoffer, Tenn.	Holston	East Tenn. & Va.	300 men

is in bad situation in such that we could never redeem it but we are in high spirits yet, and still look forward to the day of our redemption, and I think it is not far off," related W. R. Lacey of the 6th Tennessee. Presstman was assigned a new officer, Lieutenant Robert L. Cobb. A Kentucky native, the twenty-three-year-old was serving as the civil engineer for the city of Clarksville, Tennessee, at the start of the war. He enlisted as a private in the 50th Tennessee but was quickly assigned to ordnance duty. Cobb was at Fort Donelson but managed to avoid capture, briefly serving in his cousin's Kentucky battery. It took over a year for him to be properly assigned to the engineers, and he would not report for duty until the summer of 1863.[7]

On January 26, 1863, Hardee dispatched Captain Green and Lieutenant Helm to Bragg's headquarters with a map of the Tullahoma vicinity. "The engineers will explain whatever the map may fail to show you, respecting the topography of the ground it is proposed to fortify," he wrote. Hardee requested that the army commander show them the points to be fortified and the nature and extent of the fortifications. The map revealed few defensive advantages, Hardee continued, and he warned that the town could easily be flanked. Meanwhile, "Bragg's engineer," the specific one was never named, advised the army commander that significant bridge repair needed to be done on the Tennessee & Alabama Railroad between Columbia and Athens, Tennessee. The message was passed on to Joseph E. Johnston, the theater commander, but despite repeated dispatches from Gilmer in Richmond, no action was ever taken.[8]

Wampler returned from his sick leave on February 8, 1863. He encountered Presstman at Bridgeport, who had been inspecting the fortifications at that place. Wampler learned that the Virginian had only days earlier replaced him as army chief engineer. Although Wampler had only been acting chief, the change came unexpectedly. That summer he requested a transfer to the Army of Northern Virginia. His stated reason was purely personal—he had been away from his family for two years and the enemy had "devastated my home, destroyed and stolen everything, abused my wife, mother, and five little helpless children." It is difficult not to conclude that being passed over for the top engineer command left him rankled. Gilmer approved the transfer, contingent upon Bragg's approval, which was not forthcoming.[9]

In mid-March, Wampler was ordered to Columbia, Tennessee, to oversee the construction of a bridge over the Duck River. Meeting on the 15th with Major General Earl Van Dorn, commanding cavalry on the army's left flank, the engineer learned that the bridge was not far enough advanced to make an inspection. He nonetheless scouted the area and, with Bragg's subsequent approval, repositioned the project twelve miles, to the west of Columbia. It was never determined who destroyed the 225-foot Hickory Creek Bridge, on a spur of the Nashville & Chattanooga Railroad between Tullahoma to Manchester, on the night of April 7. Wampler was ordered to take a fifty-man detail and repair it. A delay in obtaining supplies and torrential downpours slowed the project, which was not completed until the 21st. Wampler had barely returned to Tullahoma before he was ordered to Bridgeport to inspect the fortifications. Once back in Tullahoma, he and Presstman were sent to scout potential bridge sites on the Duck River. After fighting a bout of dysentery, Wampler, on May 20, was inspecting the Elk River bridge fortifications at Allisonia. He next traveled to Wartrace, Tennessee, to map the topography on Hardee's front, and then made a reconnaissance to within fifteen miles of Murfreesboro. It was the same for Lieutenant Helm. From April 25 to May 8 he superintended bridge construction over the Duck River. From June 5 to June 18 he made reconnaissance missions around Shelbyville and Bell Buckle, and did the same on the 20th between Fairfield and Riley's Ford. Such was the life of an engineer.[10]

Although Bragg's army occupied a wide front in the spring of 1863, his plan was to concentrate and make a stand at Tullahoma if forced to withdraw. Preparing the town's defenses became an engineering challenge. The works comprised a semi-circle with six redoubts along the Military Road west of Rock Creek and another seven redoubts north and east of town that stretched from the Manchester Road to the Fayetteville Road. Abatis fronted the works, and all the trees and houses were cleared for 1,500 yards. Fort Rains, a one-acre bastion with a dozen guns and capable of holding 500 infantry, was positioned on the railroad northeast of town. An eight-foot-deep, twelve-foot-wide trench surrounded the fort. In March, William Preston Johnston inspected the works and expressed criticism: "[M]uch labor has been wasted on them," adding that they were "too weak to rely upon and too strong to abandon to the enemy." The "slight redoubts" were typically not flanked by rifle pits. He saw little advantage in the ground elevation, although Wampler assured him that one existed. Bragg related to him that extensive earthworks had a tendency to demoralize the troops, and that he would go on the offensive. If such was the case, observed Johnston, the abatis would be an obstruction. Sayers, on June 12, requested 1,000 men from Polk's Corps to report to Pharr to extend the fortifications on the east side. Sayers and Morris inspected the lines on the 29th. In their absence, Presstman ordered most of Polk's tools to Hardee's Corps. A miffed Sayers requested that they be promptly returned.[11]

When the rebels later abandoned the works, Federal engineers were able to make their own assessment. Fort Rains was classified as a "bastioned earthwork of four points without works." The approaches were well-covered, all of the timber having been cut down to within musket range. The Elk River bridge, three miles to the south, had a stockade on the north side of the river and a small circular work at a ford 200 yards distant. On the south side of the river were two square works joined at the corner. Some timber was still standing on both sides of the river; indeed, the timber on the north bank closely approached the bridge. No defensive works had been built at Estill Springs, nor surprisingly at the vital Cowan tunnel.[12]

Wampler and Lieutenant Thomas Newcomb of the engineers arrived at Bridgeport on April 23 to make an inspection of the fortifications. The works protecting the vital Tennessee River bridge comprised a semi-circle along the high ground on the west bank. Wampler found Batteries 1, 3, and 5 well-constructed with good drainage, but the angles were "too lightly covered with earth." The main concern was the so-called "Hill A," opposite Battery 5 bordering the river. It had great elevation but was simply too far outside the fortified perimeter to extend the line. The only alternative was to occupy it with infantry.[13]

Presstman meanwhile continued to struggle with an undermanned engineer staff. In March 1863 there were at Army of Tennessee headquarters only three other engineers besides himself—Wampler, Buchanan, and McFall, with Conrad Meister as draftsman. Hardee's Corps included Green, Helm, John F. Steele, Force, and Rowley. Polk's Corps had Sayers, Pharr, and Morris. There were two engineers with the cavalry—Silvanus Steele with Wheeler's division and Amos S. Darrow with John Morgan's division. These thirteen civil engineers, none of them West Point graduates, plus the two sapper companies, comprised the entire engineer corps of the South's second largest army.[14]

Presstman spent much of his time during April and May requesting more of everything—from entrenching tools, T squares, compasses, and levels to lumber, rope, and drawing supplies. On June 1, Gilmer replied that only 400 of the requested 1,000 axes could be sent. The next month he did send the requested 1,000 shovels. Two shipments received at Chattanooga, on August 20 and 31, revealed the wide variety of tools and equipment required. Obtaining additional personnel proved more challenging, with Gilmer denying repeated requests: "No additional engineer officers can be spared for Gen'l Bragg's army at this time." McFall and Thomas were dispatched to East Tennessee to inspect the Virginia & Georgia Railroad. Sayers had his own problems. On the morning of June 29, he and Morris inspected the Tullahoma trenches. The rifle pits should have been dug as soon as they were lined off, "but it is too late to do this work now," they reported. They also noted that timber should have been cut back 300–500 yards more to the north and east.[15]

During June and July 1863, Morgan, with 2,500 cavalry, raided deep into Kentucky, Indiana, and Ohio as a way of siphoning off Federal troops from Vicksburg. Although his unit destroyed bridges and created panic among northern citizens, the strategic plan failed to materialize and Morgan, with most of his men, was ultimately captured. The attached engineer, Amos S. Darrow, managed to escape and somehow made it to New York City, where his brother lived, only to be captured on August 10, 1864.[16]

A number of engineer officers sought promotion during the summer, including thirty-six-year-old Lieutenant John F. Steele. As a civil engineer in Alabama prior to the war, he had worked with the Memphis & Charleston Railroad, Tennessee & Alabama Railroad, and Central Railroad of Alabama. He graduated from the University of Virginia and during the early months of the war served in the engineer corps in that state. Stationed at Tullahoma in June 1863, he immodestly wrote Gilmer: "My experience is perhaps greater than that of any officer in this Corps, both as Military and Civil engineer." He had several officers write letters on his behalf, which was not received well in Richmond. Gilmer responded: "It would not be wise to have letters sent to the chief of the department."[17]

Others continued to request promotions, but the bureaucratic wheels turned slowly. Lieutenant Pharr had letters of recommendation from Presstman, Sayers, Wampler, Green, and Buchanan. Even the home owner in Shelbyville where Sayers and Pharr roomed sent a supportive letter. Gilmer's response closed the door: "There being no vacancy in the Corps, below the grade of field officer, all that can be done is to place the letters on file for consideration should a vacancy occur." Helm, likewise, had recommendations from both Hardee and Wampler, but on July 5 he wrote the corps commander: "I have heard nothing." Gilmer eventually submitted his nominations for promotion. Of the sixty-five available positions, only six were serving in Tennessee, and most of those in East Tennessee. Of the six vacancies for staff officers, the

Table 4.2. Shipments Received, Engineer Department, Army of Tennessee, August 20 and 31, 1863

TOOLS	658 axes, 2 sledge hammers, 4 cut saws, 257 augers, 1 tenant saw, 225 shovels, 1,738 picks, 492 ax handles, 8 hatchets, 2 foot augers
ENGINEER INSTRUMENTS	8 rules, 8 framing squares, 13 planes, 8 compasses, 1 T square, 3 drawing knives, 5 steel squares
ADMINISTRATION	12 company books, 300 envelopes
CONSTRUCTION MATERIALS AND INSTRUMENTS	28 chisels, 40 handsaw files, 1 Basel square, 12 crosscut saw files, 2 gimlets, 2 hammers, 11 oil stones, 5 tape lines, 125 lbs. rope, 6,075 ft. lumber, 8 claw hammers
MISCELLANEOUS ITEMS	21 packing boxes, 4 lbs. horseshoe nails, 14 lbs. steel, 18 kegs horseshoe nails and spikes, 81 empty barrels, 6 barrels rosin, 12 barrels coal oil, 3 bench saws

Heartland had no nominations. There were twenty-nine vacancies for captain, but only Richard McCalla and Arthur W. Gloster were tapped. Of the thirty openings for first lieutenant, Henry Pharr, T. S. Newcomb, Felix R. R. Smith, and Robert Cobb were selected.[18]

Upon Von Sheliha's prisoner exchange from Island No. 10, he was assigned to East Tennessee, where he became Buckner's chief of artillery and later chief of staff. Norquet was transferred to Knoxville in June. Although seemingly a demotion, the Knoxville office was actually larger than that in Chattanooga. Norquet, who impolitely took all of his office furniture and equipment with him, inherited thirteen assistant engineers, a draftsman, and a clerk. On the 7th, Wampler finally got his transfer—not to Charleston or the Army of Northern Virginia as requested, but to replace Norquet at Chattanooga. He quickly busied himself with requesting funds for large construction projects and drawing maps. Between June 20 and June 24, he was at Loudon drawing field sketches of the river crossing. McFall and Thomas meanwhile made surveys north, east, and west of Chattanooga. Wampler, on July 23, was next ordered to Atlanta to work on the city's fortifications. He was there less than two weeks before again moving, this time to Charleston; he was at last back with Beauregard. Ten days after his arrival, while at Fort Wagner, he was killed by the explosion of a Union naval shell.[19]

Mississippi and Alabama

Following the surrender of Vicksburg (to be discussed in chapter 5), Joseph E. Johnston's Army of Mississippi withdrew to Morton, Mississippi. The reinforcements arriving from the Atlantic coast had no wagons, mules, or artillery. It would not be until the end of June that wagons and mules could be supplied, mostly from Georgia, and a dozen field guns mounted on carriages made at Canton. The army at that time totaled one cavalry and four infantry divisions, with 26,300 troops.[20]

The engineer corps was deprived of the services of several officers following the surrender of Vicksburg. James M. Couper was reassigned to Savannah following his prisoner exchange, and John G. Kelly, serving as Louis Hebert's assistant adjutant general, went to Wilmington, North Carolina. Leon Fremaux remained so desperately ill at the surrender of Port Hudson that the Federals did not even bother to parole him. He nonetheless recovered and made it through the lines to Mobile. Lockett was not reassigned until late October 1863, when he arrived at Meridian, Mississippi, continuing in his role as department chief engineer.[21]

Hardee was reassigned to Mississippi, arriving on July 19, 1863; he did not like what he saw. The Army of Mississippi was reduced by desertions and somewhat demoralized. Furthermore, he wrote Polk, "I don't think Johnston knows exactly what to do with me. He seemed glad to see me, but my status is undetermined." Hardee was subsequently assigned to reorganize the Vicksburg parolees at Enterprise, Mississippi, and Demopolis, Alabama. His chief engineer, John W. Green, was ordered to Pascagoula, south of Meridian, to make a reconnaissance between that place and Mobile, to determine how Federal communications could be disrupted if Mobile was ever besieged.[22]

Gilmer wrote Johnston on August 18 that a temporary replacement bridge over the Pearl River was badly needed. "Major Meriwether, Captain [J. A.] Porter, and other engineers, have much experience as civil engineers, and can direct all operations," he wrote. The project was given such a high priority that Captain Grant, then heavily engaged in Atlanta's defenses, was sent to oversee initial operations. "Captain Grant is a civil engineer of twenty-five or thirty years' experience, and of high standing in his profession," Gilmer concluded.[23]

In late November, Captain Powhatton Robinson at the engineer office at Meridian directed Lieutenant Ginder to make a reconnaissance between that place and Dalton, Georgia, noting the topography, distances, bridges, ferries, mills, churches, and residences. Joining him in the project would be twenty-nine-year-old Captain John Baptist Vinet, a former New Orleans businessman who had been captured at the fall of the city. Subsequently exchanged, he had been engaged on building a bridge across the Tombigbee River.[24]

In February 1864, William T. Sherman entered Jackson, Mississippi, for the third time in less than a year. His forces continued west toward Meridian and, Confederate authorities feared, a potential move against Mobile from the north. Loring's division in Meridian had no choice but to retreat, while W. A. C. Jones's sapper company destroyed bridges and obstructed roads. Polk withdrew Loring into Alabama—"I scared the Bishop out of his senses," wrote Sherman. He then began the work of destroying miles of the Southern and Mobile & Ohio railroads. As for Meridian, the town "no longer exists," reported Sherman, but no move was made on Mobile. Polk began strengthening his Alabama defenses, specifically around Demopolis and along the Tombigbee River. While Porter and his sapper company worked at Demopolis, Lieutenant Richard A. O'Hea oversaw operations at Bluff Port. The forty-three-year-old Irish emigrant had previously been detailed to the engineers from the 22nd Mississippi. George Donnellan was in charge at Hay's Ferry, and Lieutenant S. McD. Vernon, who had served under Fremaux at Port Hudson, Louisiana, oversaw operations at Jones Bluff.[25]

Lemuel P. Grant

Lemuel P. "L. P." Grant was born in Frankfort, Maine, in 1817. A civil engineer by profession, he came to Atlanta in 1849 to survey for the railroad. He was instrumental in the construction of the Western & Atlantic and Georgia Railroads, and he later served as superintendent of the Montgomery & West Point Railroad. Grant purchased large tracts of land in what became southeast Atlanta, and he built an impressive two-story stone and brick house that survives to this day. In profession and gravitas, Grant was the right man at the right time.[26]

Although best known as the engineer who designed the Atlanta defenses, Grant, heading the Atlanta engineer office, was also involved in projects in North Georgia, East Tennessee, Mississippi, and northern Alabama. In Alabama he oversaw the construction of a salt manufactory in Clark County. He contracted with Amory Dexter, a thirty-four-year-old civil engineer, as the on-site professional. Originally from Massachusetts, Dexter came to Atlanta to do work for the railroad. He later became the agent for the Hydraulic Hose Mining Company in northeast Georgia. The Clark County project was shut down in early September 1863.[27]

Another project under Grant's supervision was the railroad construction between Blue Mountain (modern-day Anniston), Alabama, and Rome,

Georgia—a distance of fifty-one miles. The project was well underway before the war and by May 1862 extended from Selma to Blue Mountain. The idea was now to connect the Selma industrial complex with Chattanooga. Acquisitions of labor and the actual grading of the road was the responsibility of the state of Alabama, and the rails were to come from lesser used tracks. Grant was nonetheless called upon to inspect and administer the project.[28]

Another task that fell to the Atlanta office was the construction of Howe truss bridges. In 1840 a Massachusetts engineer, William Howe, invented a type of truss bridge that was widely in use by the Civil War. Replacement bridges (or "duplicate bridges" as they were called) were assembled in Atlanta. The timber was cut under contract south of the city and then pre-assembled by carpenters once in the city. The bolts, connecting plates, and other iron parts were made in the Atlanta machine shop of Anthony L. Maxwell (relocated from Knoxville), the Etowah Iron Works in Cartersville, Georgia, and the Tredegar plant in Richmond. "Keep a supply of bridging (Howe truss) on hand," Gilmer wrote Grant on January 17, 1863. In early April 1864, Grant reported: "Duplicate bridges are now being framed for the Western & Atlantic. I have lumber on hand for duplicate of Strawberry Plains and on East Tennessee & Virginia Railroad. In three months can have timber sufficient to duplicate (or rebuild) the structures on East Tennessee & Georgia and the Nashville & Chattanooga Railroad." In preparation for the Atlanta Campaign, still more duplicates had to be prepared. "Duplicate bridges are ready except for rods and other irons for Chattahoochee, Pettus Creek, and five other crossings of Chickamauga Creek. I shall order a duplicate of Oostanaula River. If an emergency should arise at Resaca, three spans of Chickamauga Creek Bridge can be readily adopted," Grant notified Presstman on May 12.[29]

Grant's high-priority project remained the defense of Atlanta, which served as the logistical hub for the Army of Tennessee. A Union cavalry raid by Colonel Abel Streight in northern Alabama, in the spring of 1863, was foiled by Nathan Bedford Forrest. The close call nonetheless alerted the Confederates that it was time to prepare the city's defenses. Gilmer notified Grant that summer to begin examining works for the city and the Chattahoochee River. Ever mindful of his lack of military experience, Grant appealed to Gilmer for assistance. The colonel extended his complete confidence in his abilities—his "natural gift" of judging topographical features, his specialty in the construction of roads and bridges, and his experience in preparing maps and surveys made him totally qualified for the task at hand. As for asking for help in preparing field works, rifle pits, and lines of infantry cover, Gilmer suggested that Grant would learn on the job "without any difficulty."[30]

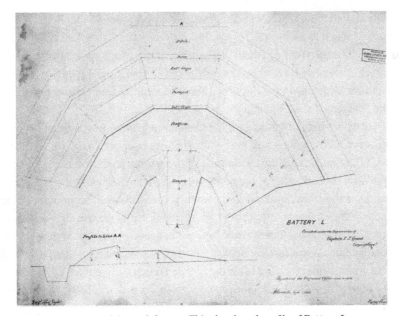

Fig. 4.2. Battery L, Atlanta defenses. This sketch and profile of Battery L was prepared in the Atlanta engineer office under the direction of Lemuel P. Grant. Courtesy Kenan Research Center at the Atlanta History Center.

On August 4, 1863, Gilmer proposed defensive works comprised of twelve to fifteen forts arranged in a 10- to 12-mile perimeter. He insisted that the works needed to extend far enough from the city to avoid Atlanta's destruction by Federal artillery. All trees fronting the works should be cleared for 900–1,000 yards, and the stumps should be cut low. By August 17, small works had been completed at the ferries north of the city and three of the city forts were under construction, although a dearth of labor hampered progress. The city defenses were well advanced by November, the work being performed by hundreds of impressed slaves. There were seventeen redoubts, generally for five guns each, connected by rifle pits. Four redoubts were incomplete at that time, and the plank platforms were yet to be laid. The works extended for seven and a half miles. Unfortunately, topographical considerations and a limited garrison meant that Grant could not expand the fortifications wide enough to prevent the city from being shelled. In his analysis of the Atlanta defenses, Robert J. Fryman concluded that the works lacked the sophistication of those of Vicksburg and Richmond. The blueprints of various forts reveal that they consisted of right-angle lines with a ten-foot-wide parapet (sufficient to absorb a 12-pounder-shot).[31]

On May 6, 1864, Grant was assigned twenty-one-year-old Lieutenant William A. Hansell to assist him. A student attending the Georgia Military Institute in Marietta, Georgia, at the beginning of the war, he initially served in the 35th Alabama. He was later transferred to the engineer department in Columbus, Georgia, under Captain Theodore Moreno, and had been working in the District of West Florida.[32]

Overwhelming demands, ranging from large bridge construction projects to massive fortifications and everything in between, overtaxed the engineer corps to the breaking point. A scarcity of everything from iron parts for trestles to shovels and axes for trenches delayed projected schedules. The huge expanse of the Heartland left large sectors vulnerable and communications tenuous. Projects could oftentimes be completed, but never in a timely fashion. Exhausted engineer officers went from enterprise to enterprise to meet the demands, but their efforts could not offset military losses on the battlefield. The army was simply trying to accomplish too much with too little. The expansive geography of the Heartland, coupled with excessive demands, was more than the engineer corps was capable of handling.

5

ENGINEERING COLOSSUS

IN MARCH 1862, Beauregard quietly conceded that Island No. 10 and Fort Pillow would only be "temporary barriers," and that additional defenses would have to be prepared south of Memphis, at Helena, Arkansas, Vicksburg, Mississippi, or Port Hudson, Louisiana. Major General Lovell, commanding at New Orleans, dispatched an officer to Jackson, Tennessee, on April 20 requesting Beauregard's assistance at Vicksburg. The next day Beauregard ordered David Harris, then at Fort Pillow, to report to the "Hill City" to make a reconnaissance for obstructing the mouth of the Yazoo River and constructing batteries along the Mississippi River above Vicksburg. His first concern was the Union flotilla on the upper Mississippi. He nonetheless saw the possibility that the Federals might cut a canal across DeSoto Point, the hairpin turn opposite Vicksburg, so Harris was also instructed to prepare a battery south of the city. As in the case of Columbus and Fort Pillow, Beauregard envisioned a small garrison of only 3,000.[1]

A Federal fleet under Admiral David Farragut sailed north from the Gulf of Mexico, passed the lower Mississippi River forts, and captured New Orleans on April 26. The episode, according to Timothy B. Smith, disclosed the primary defect of Confederate Western Theater strategy—the lack of an in-depth defense. As in the surrender of Fort Henry, which exposed the Tennessee River all the way to northern Alabama, so too the fall of New Orleans opened the lower Mississippi River for 200 miles. Vicksburg, at least at that time, remained a "soft target," and Confederate authorities now had to worry about both the lower *and* upper Federal fleets. All previous engineering projects in the West paled to the magnitude of the challenges that lay ahead.[2]

Flawed River Defenses?

Colonel James Autry was the first field officer at Vicksburg. Upon his arrival he found "not a soldier, not a gun, not a pound of powder, not an engineer. Shortly after an engineer officer [Harris] reported for duty." Due to the scant slave labor, troops were immediately put to work on the river batteries as soon as they arrived. Some twelve to thirteen guns were mounted on Autry's watch.[3]

Brigadier General Martin Luther Smith arrived on May 12 to assume command. New York-born and a graduate of West Point (class of 1842), Smith served in the topographical engineers in the Regular Army, surveying the coast and rivers of Florida and Georgia. During the Mexican War he mapped the area around Mexico City. When the US Army departed, he remained at the request of the Mexican government to devise a water drainage system. In 1853, Smith was appointed head of the US Coast Survey. He later resigned from the army and became a civil engineer for the Fernandina & Cedar Key Railroad. Tall, trim, beardless, and described as "thoroughly military," he served in the Confederate engineers for seven months at New Orleans before being promoted to colonel of the 21st Louisiana, and shortly thereafter to brigadier general. Upon his arrival at Vicksburg, only three batteries had been completed. For the next six days crews labored day and night mounting guns from Pensacola and New Orleans, completing an additional three batteries.[4]

Farragut's fleet arrived at Vicksburg on May 18, followed shortly by David Porter's mortar schooners. Days of shelling followed, destroying houses and killing several, but in the main the shells proved nearly harmless. Craters were measured at the depth of seventeen feet, making bombproofs virtually useless. Major Samuel H. Lockett later recalled the morale effect: "One of my engineer officers [Fremaux?], a gallant officer who had distinguished himself in several severe engagements, was almost unmanned whenever one passed anywhere near him. When joked about it, he was not ashamed to confess, 'I no like ze bomb; I cannot fight him back!'"[5]

On June 28, Farragut's vessels ran the Vicksburg river batteries and united with the Union fleet moving south from Memphis. Precisely how many batteries were constructed by that time is not known. In a reference to what appears to be the Water Battery and Batteries 6 and 7, a Federal eyewitness wrote: "At the upper end of the city is a three-tiered battery having one large gun in a water battery, two more half-way up the hill, and three others on the summit of the bluff." In another reference to what appears to be the Marine Hospital and Blakely Batteries, the correspondent noted that he counted nine guns in two batteries on the summit of the bluff.[6]

Fig. 5.1. A former member of the US. Topographical Engineers and later a railroad civil engineer, Martin L. Smith served in the dual role of commanding an infantry division and having engineer oversight of the construction of the river batteries. Courtesy Alabama Department of Archives and History, Q50.

Porter later commented about why initial efforts to run the batteries failed. He wrote that he "had officers stationed all along [the opposite bank] to note the places where the guns fired from, and they were quite surprised to find them firing from spots where there were no indications whatever of any guns before. The shots came from banks, gulleys [sic], from railroad depots, from clumps of bushes, and from hill tops, two hundred feet high. A better system of defense was never devised." The advantage of camouflaged batteries notwithstanding, the river batteries neither sunk nor seriously damaged any of the ships even though moving against the current doubled the vessels' exposure time. Unable to capture the city without army support, Farragut eventually withdrew.[7]

The next time the Union navy attempted to run the batteries, the defenses were far stronger. By the time the fleet—with four ironclads, transports, and support vessels—ran the gauntlet on the night of April 16, 1863, there were thirteen batteries with thirty-three guns, stretching three and two-thirds miles. The rebel gunners, although caught off guard, managed to fire over 500 rounds, scoring about seventy hits. The vessels, though battered, were not defeated and managed to get south of Vicksburg. On the night of April 19, more transports and barges made it down, though one transport was sunk.[8]

On December 20–22, 1862, newly appointed theater commander Joseph E. Johnston examined the river defenses, along with President Davis. Johnston was not pleased with what he found. The water batteries were planted to prevent the bombardment of the town, rather than to close the navigation of the river. The small number of heavy guns "had been distributed along a front of two miles, instead of being so placed that their fire might be concentrated on a single vessel." He had been told that an attack was imminent, however, meaning that "such errors could not be corrected." Confederate engineer James T. Hogane witnessed the event on April 16 and, in postwar years, expressed criticism of southern engineering. "The effect of firing on moving objects by single guns, proved itself, as it did in other instances, a failure, and confirmed the opinion that I had always held, that concentrated mass firing is the only effective way to destroy iron-clad vessels of war." He concluded that the engineer in charge of construction (Harris and M. L. Smith) "should have grouped his guns." Also, the batteries should have been located "below the skyline."[9]

Vicksburg historians Edwin C. Bearss and Warren E. Grabau concurred. They evaluated that the best way to defeat the Union fleet "would have been to group the heavy rifles into closely spaced batteries at the Water Battery. ... [T]he guns should have been placed in the best possible position to use their great range and penetrating power against the ironclads. This could best be done by placing them at the Water Battery, where the ironclads would be exposed to raking fire for a considerable distance as they rounded the tip of DeSoto Peninsula. Downstream there were only wooden ships, against which the destructive power of the big shell guns, such as the 10-inch columbiads, would have been effective. A position near the Marine Battery would have been suitable. Each battery should have had guns of the same type and caliber. This would have made fire control much easier, and would have much simplified ammunition supply."[10]

One of the difficulties facing Confederate engineers was the soil. The base of the Vicksburg bluffs was hard limestone, but the bluffs were made of loess, a yellowish, wind-blown silt. When used to construct batteries, the particles were so fine that thicker parapets had to be made. To prevent penetration, the bluff parapets were thus an incredible thirty to forty feet thick! This meant that once the Union vessels hugged the shoreline, the guns could not be sufficiently depressed. Additionally, if a gun is depressed to its maximum, the shot

Table 5.1. Vicksburg River Batteries, April 1863

BATTERY	ELEVATION*	RIFLED GUNS	HEAVY SMOOTHBORES	MEDIUM SMOOTHBORE	LIGHT RIFLES
Water	30	2	1	1	—
7	110	—	2	—	—
6	160	—	—	1	—
5	90	1	—	—	—
4	60	—	1	—	—
Wyman's Hill	40	—	4	—	2
Whig	60	—	1	1	—
Depot	100	—	1	—	—
Railroad	100	—	—	—	2
Brooke	40	1	—	—	—
Marine	40	2	—	5	—
Widow Blakley	130	4	—	—	—
South Fort	170	—	1	—	—

* Feet above mean river level

may separate from the powder charge, thus reducing accuracy. The constant loading (leveling of the barrel) and then firing (depressing the barrel) made for a tediously slow rate of fire.[11]

Grabau concluded that most of the effective Confederate batteries were not on the bluff, but were the Water, Wyman's Hill, and Marine Hospital Batteries, all thirty to forty feet above river level. Battery No. 7, at a 110-foot elevation, was so high that "the Confederates themselves complained that its great elevation made it ineffective." The Blakley rifle was subsequently moved, the bluff being "too high." Nonetheless, a Federal sailor involved in the 1862 running of the batteries noted that "the batteries on the summit of the hills cause our navy men more trouble than those lower down." Wrote another: "The rebels were driven out of every battery, except one—the battery on the summit of the bluff below the town. This battery continued firing as long as the vessels were within range." Most of the shore guns fired too high, causing little damage to Farragut's wooden vessels, but tearing the rigging to pieces. Whether or not this was due to the elevation of the batteries, caused by the placement of the engineers, or by inexperienced Confederate gunners is not known.[12]

Once the Federal fleet ran the batteries in 1863, it had to either reduce or run the fortifications at Grand Gulf. Lieutenant General John C. Pemberton, commanding at Vicksburg, had foreseen such a possibility and sent Brigadier

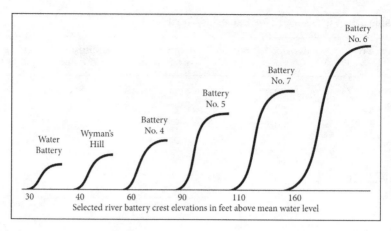

Fig. 5.2. Selected river battery crest elevations above mean water level. .Grabau, *Ninety-Eight Days*, 48–50.

General John S. Bowen with his brigade to that location. Major Lockett lined off the works and left construction supervision to Lieutenant George Donnellan of the engineers. In the brutal six-hour engagement that ensued, several of the ironclads were badly battered and the forts were heavily damaged by over 2,300 rounds, dismounting all of the guns but one. The fleet continued south, and the crossing from the Louisiana side of the river was made downriver at Bruinsburg, Mississippi.[13]

Personnel

Major Lockett was reassigned to Vicksburg on May 18, 1862. Initially he performed only the mundane task of laying out camps, but he was soon sent to construct a battery on the first bluff of the Yazoo River. He was ordered by Harris to employ the nearby sawmill to prepare lumber for the gun platforms. The transport *Magnolia* would be sent with twelve picks, twenty-five shovels, and block and tackles. Harris was returned to the Army of the Mississippi, and Lockett, now the engineer in charge, saw the completion of the river batteries. "The garrison was engaged in strengthening the batteries already constructed, in making bombproof magazines, and in mounting new guns recently arrived," recalled Lockett. "Several new batteries were laid out by myself on the most commanding points above the city; these were afterwards known as the 'Upper Batteries.'"[14]

On September 22, Lockett traveled up the Yazoo River to inspect the fortifications—"All going on steadily and slowly," he wrote. Appointed on November

Fig. 5.3. Samuel H. Lockett. A West Point graduate and former member of the US Army Corps of Engineers, Lockett superintended the construction of the massive Vicksburg land defenses. Courtesy Tennessee State Library and Archives, Carte de Viste Collection.

1 as chief engineer of the Department of Mississippi and East Louisiana, with headquarters in Jackson, he was frequently pulled away from Vicksburg, although he never relinquished responsibility for the defenses. On the 5th, he wrote his wife from Jackson: "I am going up to Holly Springs this evening and will probably be absent several days. General Pemberton wants some fortifications made on the Hatchie River, some twenty miles this side of Holly Springs. As soon as I get things started there[,] I will return I suppose to this place unless Genl. P. takes the field in Person when I may have to go with him." On the 25th he was examining the works at Port Hudson. The town reminded him of Fernandina—"old and quite dilapidated." By December 17 he was at Grenada.[15]

Frustrated over his lack of rank, on February 13, 1863, the major addressed the issue with Gilmer; he did not hold back. Others had been advanced over him and he felt that he "had been somewhat unjustly treated," he wrote, not knowing exactly if he should take it personally, which he clearly did. "My patience is now about exhausted," he continued, concluding in disgust, "I suppose the whole affair has been settled by being 'pigeon holed.'" Lockett, perhaps unknown to him, had in fact already been recommended for promotion to lieutenant colonel the previous November.[16]

In early December 1862, Captain I. J. Thysseus arrived in Vicksburg. Formerly of the French cavalry and having studied at Saint-Cyr, he immigrated to the United States about 1846 and taught military tactics and engineering at the University of Nashville. He had previously served as engineer with Major General Earl Van Dorn in Arkansas. While at Vicksburg he was the engineer-in-charge in Lockett's absence. When Lieutenant William W. Fergusson reported as a mapmaker to Lockett's office, Lockett "needed field engineers as well as draughtmen, [H. A.] Pattison being the only one in charge of the office and Thysseus unable to ride and by no means up with the maps to be copied." Thysseus was not present during the subsequent siege, and he is believed to have returned to Van Dorn's staff until the general was murdered in the spring of 1863.[17]

Of the eighteen officers serving on Lockett's engineer staff during the subsequent siege of Vicksburg (one for each brigade and division), most were merely detailed as engineers. Only five officers actually had professional engineering training. James M. Couper was a West Point graduate (class of 1856) and a student of Mahan. Indeed, the instructor had given him a letter of recommendation stating that he was "well versed in topographical and other engineering drawing and the use of instruments for surveying levelling." Another officer described him as "an educated engineer and accomplished draughtsman." His map of Fort Pemberton on the Yazoo River has unfortunately not survived.[18]

Thirty-year-old George Donnellan served as a civil engineer in the Midwest prior to the war, performing jobs for the Mount Pleasant & Muscatine Railroad in Iowa and the Wabash & Mississippi River Railroad in Illinois. He was a virtually unknown in Virginia at the start of the war and became part of a Washington, D.C., spy ring, in which he passed information to Beauregard and his chief of staff, Thomas Jordan. The ring was eventually broken up and arrests were made, but Donnellan managed to escape. He eventually came to the West, apparently through Beauregard's influence, and ended up at Vicksburg. Lockett's after action report mentioned him as being "in charge of procuring and distribution of materials." Donnellan died on December 3, 1865, whether by disease or some other cause is not known, and was buried in Cedar Hill Cemetery in Vicksburg.[19]

Powhatton Robinson of Mississippi, described as a "practical and scientific civil engineer," initially served in Virginia but was reassigned to Vicksburg in October 1862. He was the engineer in charge of constructing a cotton bale fort covered with dirt (Fort Pemberton) on the 500-yard-wide neck separating the Tallahatchie and Yazoo Rivers near Greenwood. He requested to be relieved of duty on March 15, 1863, due to ill health, but he continued in the engineers after the surrender of Vicksburg. The remaining engineers were Arthur W. Gloster, who, as previously mentioned, was a thirty-year-old former Tennessee civil engineer, and H. A. Pattison, a civil engineer serving as a draftsman, and whose February 1863 map of the Vicksburg defenses appears in the *Official Atlas*.[20]

There are no surviving records for draftsmen R. R. Southard, whose records show him as "citizen acting 1st Lt. Engineers." James T. Hogane was a surveyor living in Davenport, Iowa, at the start of the war. He was captured at New Madrid in March 1862, exchanged, and arrived in Vicksburg in mid-April 1863. J. J. Conway was detailed from the 20th Mississippi, and after his prisoner exchange commanded a company of engineer troops in the Trans-Mississippi. Thirty-year-old Henry Ginder, born in Kentucky to German parents, had served in the US Coast Survey and possessed remarkable drawing skills. Prior to the war he went into partnership and owned a New Orleans jewelry store. He served in the Crescent Regiment and later in Fenner's Louisiana Battery. "There was a need of surveyors and Engineers & as I had experience in that line," Ginder recalled, he was detailed to Captain Fremaux at Port Hudson. He was later ordered to Vicksburg to serve as a draftsman.[21]

Lockett made mention of "Colonel D. H. Huyett" of the engineers who carried important papers to Richmond. Thirty-one-year-old Huyett had the title before the war, and it is inscribed on his tombstone in Shelby County, Alabama, but there is no indication he ever served in the Indian or Mexican Wars, much less as a colonel. His Confederate record officially lists him as a citizen from Mississippi under contract with the engineers. It seems unlikely he served in a state militia; perhaps his rank was honorary. Before the war he, along with Nathan H. Parker, published *The Illustrated Miner's Handbook & Guide to Pike's Peak,* outlining trails in Kansas and western Nebraska. Both before and after the war, Huyett provided illustrations for *Frank Leslie's Illustrated*. It seems reasonable to assume that, beginning in January 1863, he drew maps for the engineers. Concerning Lockett's statement, a pay voucher indicates: "For actual expenses incurred in travelling as special messenger from Vicksburg to Richmond with important papers and maps, between 17th May and 13th June 1863."[22]

Fremaux served as chief engineer at Port Hudson. His October 30, 1862, map of the defenses, drawn at a scale of four inches to a mile, was beautifully prepared and survives in the Gilmer Papers. Important roads were drawn

Table 5.2. Engineer Personnel, Department of Mississippi and North Louisiana

OFFICER	POSITION	COMMENT
Maj. Samuel H. Lockett*	Chief engineer	West Point
Lt. George Donnellan	Engineer	Former civil engineer
Henry Ginder	Civilian	Draftsman
G. C. Bower	Civilian	Clerk
Capt. Powhatton Robinson	Assistant engineer	Former civil engineer
Capt. J. J. Conway	Assistant engineer	
Lt. Arthur W. Gloster*	Assistant engineer	Former civil engineer
Lt. R. R. Southard	Assistant engineer	Draftsman
Capt. James M. Couper	Assistant engineer	West Point
B. A. Sanders	Civilian	Office assistant
Capt. Deitrich Wintter*	Former architect	
Lt. E. McMahon*	Sapper Company	Former machinist
Lt. Francis Gillooly	Sapper Company	
Capt. James Hogane	Assistant engineer	Former surveyor
Lt. S. W. McD. Vernon	Assistant engineer	
Lt. P. J. Blessing	Assistant engineer	
OTHER ASSIGNMENTS		
Maj. Minor Meriwether*	North Mississippi	Former civil engineer
Capt. Leon J. Fremaux*	Port Hudson	Former civil engineer
Lt. Frederick Y. Dabney	Port Hudson	Former civil engineer
D. H. Huyett	Civilian	
Capt. John G. Kelly	Louis Hebert's AIG	Former civil engineer
Capt. Jules V. Gallimand	Sapper Company	Former chemist

*Also served in the Army of the Mississippi/Army of Tennessee

in brown and local roads in red. A table included measurements of ranges from the various batteries to specific locations. Batteries 1–7 were constructed along the bluff at a range of 85 feet above river level, with Batteries 8–11 ranging from 31.9–57.5 feet.[23]

In late August 1862, Leon Fremaux was ordered to Port Hudson to make an examination of topography of the levee bordering the Mississippi River, from Robinson's plantation to Trudeau's store, with the prospect of planting a light battery of 24-pounders. "The levee is a good protection and a range of not over 300 yards," he reported. Once completed, he began the larger task of lining off larger batteries for heavy guns, "as well as the line of breastworks in the rear and several redoubts, at the principal roads entering Port Hudson." Fremaux remained at Port Hudson throughout the siege. Assisting him would be Lieutenant Frederick Y. Dabney, twenty-eight, a Mississippi civil engineer previously engaged in railroad work.[24]

In January 1863 Captain Jules V. Gillimard's New Orleans company of sappers and miners, with thirty-three men, arrived in Vicksburg to serve arsenal duty. The outfit had previously worked at Fort Pillow and had performed road and bridge repair in North Mississippi. The captain, who listed his occupation as "chemist," had been instructing his men in saps and fortifications, with classes in mines, bridges, pontoons, and reconnaissance yet to come. The company was subsequently sent to near Port Hudson. In an attempt to cut off Colonel Benjamin H. Grierson's raid, Gillimard's men constructed a temporary bridge across the Amite River, but the raiders nonetheless escaped. The company remained in the Clinton-Osyka, Louisiana, vicinity until July 13, when it was dispatched to Mobile.[25]

Thirty-four-year-old John G. Kelly, born in Dublin, Ireland, lost both of his parents in infancy. Raised by his uncle, he studied as an apprentice civil engineer. At age eighteen he immigrated to the United States, where he lived in Missouri and worked as a civil engineer and surveyor for the Missouri, Pacific, & Iron Mountain Railroad. He subsequently worked for several railroads in Illinois, ultimately moving to Vicksburg, where he was employed by the Vicksburg & Ship Island Railroad. At the start of the war he raised a company of cavalry, but he was later transferred to the engineers, where he worked on Fort Thomas at New Madrid. Kelly was later on the staff of Brigadier General Louis Hebert as assistant adjutant general, but in April 1863 he was transferred back to the engineers. He examined the "crooked and narrow stream [Yazoo River], very much obstructed from the overhanging timber, the number drift piles, and cypress brakes." Lockett does not mention him in his after action report, probably because he was returned to Hebert's staff, although it is known (through a northern source) that he performed as an engineer during the siege.[26]

In his study of Union engineering during the siege of Vicksburg, Justin Solonick determined that the Federals were nearly as short of trained personnel as were the Confederates. The northerners had four members of the US Corps of Engineers, as compared to three for the Confederates. Sherman also had four West Point graduates who served as engineers at Vicksburg, as compared to one for Pemberton. The Federal Army of the Tennessee also had twelve personnel (draftsmen, civil engineers, surveyors, mechanics, etc.) who were detailed to oversee engineering tasks, the southerners thirteen. In sum, the North had nineteen engineer personnel, the South sixteen.[27]

Land Defenses

When General Joseph E. Johnston was appointed Western Theater commander in December 1862, his jurisdiction embraced both Mississippi and Tennessee. As with Beauregard, the issue of static versus mobile defense once again came to the fore. Lieutenant General John C. Pemberton, commanding field forces in Mississippi, initially placed his army behind earthworks at Grenada, with a garrison force defending Vicksburg. The works at Grenada, according to Johnston, were "so extensive that it is fortunate for us that General Grant was prevented [by a raid] from trying its strength."[28]

Initially Lockett spent a month mapping the environs east of Vicksburg. "No greater topographical puzzle was ever presented to an engineer," he recalled. The irregular line of eroded ridges and hills, coupled with thick cane undergrowth and magnolia trees, made it almost impossible even to form a general line. In their after action report, Federal engineers Frederick Prime and Cyrus Comstock remarked that there was little remaining of the eroded plateau east of the city, but "an intricate network of ravines and ridges." The sides of some of these ravines were so steep that it would require both hands to scale them. The thick growth of trees had been cut down, forming an abatis that made the terrain virtually impassable. Construction was begun around September 1 and was to some degree finished by the end of 1862. Works were also begun on Haynes's Bluff and at Warrenton, south of Vicksburg. The workforce was pathetically small: Wintter's company of sappers and miners with twenty-six men, four overseers, seventy-two hired Blacks (twenty being sick), along with three four-mule teams and twenty-five yoke of oxen. Unfortunately, the defenses were subject to erosion and had to be constantly repaired. The work was done by fatigue details, but Lockett complained that there was never more than 500 entrenching tools of all kinds.[29]

The defense line was anchored on Fort Hill, just above the Water Battery, built on seventy-one-year-old Fort Nogales, which had been constructed by the Spanish. From there the defenses ran for nearly a mile and a half on a curved east-west line on a ridge along Fort Hill Road to the Graveyard Road, which entered the city from the northeast. Near the Graveyard Road the line abruptly turned south. This angle would be a vulnerable sector, so Lockett built three works, or what Smith called "a major system of fortifications." The 27th Louisiana Lunette was located north of the road, the Stockade Redan, at the angle and slightly south of the road, and seventy-five yards south of Green's Redan. The Graveyard Road was obstructed by eight-foot-high pointed logs, arranged similar to a frontier fort. As the line (facing east) extended south,

Fig. 5.4. Square Fort. This photograph of Fort Garrett, the so-called Square Fort, was taken in the early 1920s. The structure survives today and shows remarkably little deterioration. Courtesy Archives and Records Division, Mississippi Department of Archives and History, series 0573.

the Jackson Road was protected by the 3rd Louisiana Redan to the north and the Great Redoubt to the south. The 2nd Texas Lunette protected the Baldwin Ferry Road, and the huge Railroad Redoubt was on the south side of the Southern Railroad. The Square Fort stood a half mile to the southwest. The Salient Work protected the east side of the Hall's Ferry Road.[30]

On the face of it, Lockett appears to have imprudently over-extended his line past the Salient Work 1,800 yards to the southwest past Stout's Creek to what became South Fort. This meant that two additional brigades were required to man the lines. It could perhaps be argued that he should have bent his line back at the Salient Work and constructed an east-west line, similar to the north wing, that extended to the Marine Hospital. Such a configuration would have shortened the line and freed up a brigade. Vicksburg historian Terry Winschel convincingly argued, however, that the topography simply prevented such an alteration. By extending the line farther south along the narrow ridge, Lockett was able to "traverse the valley of Stout's Bayou, thus enabling him to connect the Hall's Ferry and Warrenton Roads." Lockett's south line, as laid out, was much like a V. "Between the arms of the V is the

Table 5.3. Lockett's Defensive Line

LOCATION	AKA	PARAPET WIDTH (FT.)
27th Louisiana Redan		6
Stockade Redan		16
Green's Redan		4.5
3rd Louisiana Redan	Fort Hill	24
Great Redan	21st Louisiana Redoubt	13
2nd Texas Lunette		4.5
Railroad Redoubt	Fort Pettus	24
Square Fort	Fort Garrett	26

deep valley of Stout's Bayou which drains to the south, thus splitting the Confederate line near the base [apex] of the V. The split was obstructed by a deep abatis. Even so, Federal batteries positioned on the Hall's Ferry Road south of the Salient Work were able to fire into the rear of Confederate river batteries along the Warrenton Road [present-day Washington Street]. Federal batteries positioned along and just east of the Warrenton Road were able to enfilade that stretch of the Confederate line astride Hall's Ferry Road."[31]

Johnston, of course, was not pleased with the sheer size of the Vicksburg defenses from the outset. His initial inspection found that the works "were then very extensive, but slight—the usual defect of Confederate engineering." He added: "An immense intrenched camp, requiring an army to hold it, had been made, instead of a fort requiring only a small garrison." When Davis visited Vicksburg in December 1862, he conferred with Johnston and Pemberton. It soon became apparent, recalled Johnston, that "General Pemberton and I advocated opposite modes of warfare," that is, static versus mobile.[32]

There was, however, an alternative—make the line of the Yazoo River to the north and Big Black River to the east and south the defensive shield. Once the Federals had crossed the Mississippi River south of Vicksburg, concluded Michael Ballard, "the Confederates could use the river [Big Black] as a natural defensive barrier and hope to block any attempt by Grant to approach Vicksburg from the south. . . . If Pemberton posted his army along the high west bank bluff of the river to meet the enemy, Grant might have considerable trouble getting into Vicksburg."[33]

To some degree, Pemberton attempted such a maneuver. Once Grant landed his army at Bruinsburg, Pemberton established a defensive ring on the west bank of the Big Black River, with a brigade each at Hankinson's,

Fig. 5.5. The Railroad Redoubt. This photograph, taken in the early 1920s, shows the Railroad Redoubt still very much intact atop a prominent hill, with the Southern Railroad visible at the lower right. Courtesy Archives and Records Division, Mississippi Department of Archives and History, series 0573.

Hall's, and Baldwin's Ferries. There were, nonetheless, issues that could not be resolved. There were two topographical studies done of the Big Black, the first by Lockett in February 1863. His purpose was to find high ground where a battery could be constructed should a Federal gunboat ascend the river. At Baldwin's Ferry the high ground (fifteen feet above water level) was on the *east* bank. A mound on the bluff, another twelve to fifteen feet higher, made it the perfect location for a battery. At Hall's Ferry the commanding ground was also on the *east* bank. The only advantage was that the river bottoms overflowed to such an extent that it would be difficult for the Federals to land troops anywhere along the banks of the river.[34]

The other study was done on May 9, 1863, by Captain Henry De Veuve, an engineer on the staff of Major General W. W. Loring's division. Thirty-two years old and a graduate of West Point (class of 1852), De Veuve initially worked in the adjutant general's office in the Regular Army, but he was eventually assigned to engineer duty in Texas. At the start of the current conflict, he raised and trained several companies of infantry, but he had to resign due to ill health. He was in New Orleans at the time of the city's capture, but later managed to escape. De Veuve was assigned to ordnance duty in January 1863 but subsequently transferred to the engineers. He found the terrain between Hall's and Baldwin's Ferries "low and nearly level" for almost a mile. A bluff, forty to a hundred feet above river level, on the *west* side of the river extended north overlooking the Southern Railroad, thus giving the defender a distinct advantage.[35]

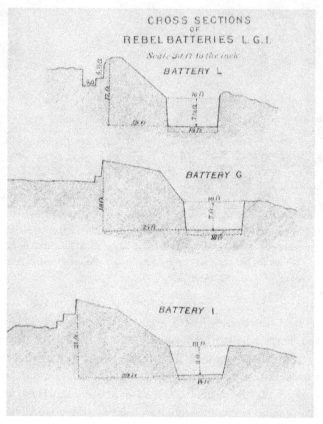

Fig. 5.6. Cross section of Confederate land batteries. These cross section profiles were drawn by Federal engineers after the surrender of Vicksburg. Davis et al., *Official Military Atlas of the Civil War*, plate XXXVI.1.

When Grant moved his army northeast toward Jackson, Pemberton sent John Gregg's brigade to delay the advance at Raymond. Johnston with 6,000 reinforcements offered only a feeble resistance in the Battle of Jackson on May 14. His forces escaped over two bridges across the Pearl River, previously constructed at the ferries by Lockett. The theater commander ordered Pemberton to move on Clinton. Rather than obey the order, the army commander put the question to his generals. Clearly, he desired to make his stand on the west bank of the Big Black, where he had a defensive advantage, but his generals urged him to move beyond the river into more flat and open ground, where his army could maneuver. Weak and indecisive, Pemberton acquiesced.[36]

Defeated at the Battles of Champion Hill (May 16) and Big Black River Bridge (May 17), Pemberton abandoned the Yazoo River line and withdrew his army into the city defenses, although he expressed to Lockett that he did not think his troops would stand the shock of an attack. He ignored Johnston's May 18, 1863, order to abandon the city. Davis, of course, had given unequivocal orders to hold Vicksburg at all hazards. The die had been cast.

Federal Attacks

As Pemberton's demoralized troops trudged back to Vicksburg, Lockett went to work on the land defenses. Since the earthworks had been constructed six months earlier and never occupied, they were now washed and eroded. Fatigue parties removed all siege pieces and field guns on the river front to the land defenses and hurriedly prepared platforms and embrasures. Portions of the works were protected by fallen trees, abatis, and telegraph wire entanglements. While inspecting the northern defenses, Captain James T. Hogane and another engineer officer discovered that M. L. Smith's division, which had not been previously engaged, occupied the rifle trenches of Indian Mound Ridge, north of Mint Spring Bayou and outside the main (but incomplete) northern city perimeter. On the night of May 18 Smith withdrew his troops. Fort Hill Ridge comprised only a stockade fort and three lunettes. "I rode along the line, staking out in hurried fashion the line of rifle pits, telling the men we would rectify mistakes at another time," recalled Hogane. Tools came down from the abandoned Yazoo River line, and fatigue parties got to work. The next day the bluecoats invested the city.[37]

Hoping to avoid a siege, Grant decided to attack on May 19, the target being the massive fortified complex at and on either side of Stockade Redan. Frank Blair's division of William T. Sherman's XV Corps took the brunt of the action. The troops enthusiastically attacked, but soon encountered thick abatis and wretched terrain. "[W]hen we got to the fort we could not climb it[.] the fort is 15 feet high and a big ditch on this side 5 feet deep & 7 wide," wrote an Illinois soldier. Few made it as far as the ditch. Total Union casualties amounted to 942; the Confederates sustained 70. Brigadier General Francis Shoup later characterized the attack as "a gratuitous slaughter."[38]

The May 19 assault, followed by the relentless Yankee shelling and skirmishing of May 20–21, revealed that the defenses, though formidable, were far from impervious. A battery of Federal 20-pounder Parrott rifles shot completely through the fourteen-foot-thick base of the Stockade Redan, although the thickness was considered standard for the day. Confederate engineers responded by adding cotton bales covered with two feet of dirt. This stopped the shells, but the bales ignited, releasing smoldering smoke. The

Confederates attempted to keep everything wet, a near impossible task in the blistering Mississippi sun and considering the natural resistance of bales to absorb water.[39]

Shoup, whose brigade assisted in holding the fortification complex at the northern angle, wrote: "Our artillery is almost useless, since we have no properly constructed protection. Being almost without intrenching tools, we can do little to repair the evil." The 2nd Texas Lunette, on the south side of the Baldwin Ferry Road, exposed many problems. "Irregularities of the ground appeared to protect the enemy as effectively as we were sheltered by our breastworks," observed Colonel Ashbel Smith. His work had no left flank protection, which meant that the exterior was exposed to an enfilade and reverse fire from the enemy approaching from the valley on the left. The interior wall was four and a half feet high, requiring the men to dig an additional two feet so that they could stand erect when they fired. They also had to strengthen the thickness of the trenches. Shells came directly through the embrasure, disabling both guns, and it proved impossible to replace them under the heavy fire.[40]

On May 22, Grant assaulted along a 3.75-mile-long section of the Confederate defenses with an estimated 40,000 troops, against 14,000 rebels. The *Chicago Tribune* declared the engagement "disastrous." A Federal described the earthworks as being at "an altitude of from ten to fifty feet, with sides in this direction, sloping and precipitous." Union casualties totaled 3,199; the southerners 500. Lockett's works, according to Smith, "had stood the test against two successive assaults." Hess credited the defeat to "Lockett's line of earthworks combined with the willingness to defend them," and the rugged terrain, which "inhibited Grant's ability to mass his men."[41]

Siege

As the Federals inched southward and occupied the Warrenton Road, Lockett became increasingly concerned that they would penetrate the flat sector between the bluffs and the Mississippi River, south of the Marine Hospital Battery. On May 25 he sent Lieutenant Donnellan and Mr. Ginder with a fatigue party to cut abatis between the river and the bluff. The protecting unit of the 41st Georgia became involved in a skirmish and the operation was cancelled, but not before 114 men of the 46th Illinois were bagged. Later that day, a trench was dug in lieu of abatis. Meanwhile, along the land defenses "the engineers were engaged in general repairs, strengthening the parapets, extending the rifle pits, placing obstructions in front of the exposed points," wrote Lockett. Sandbags were made from old tent cloth.[42]

Fig. 5.7. The Vicksburg terrain. In this photograph, taken after the surrender of the city and labeled "Behind the Confederate Lines," the barren and bleak landscape, scared with deep ravines, can clearly be seen. Courtesy Library of Congress, DC-DIG-ppmsca-35290.

Following Grant's two defeats, formal siege operations were undertaken. Enemy trenches, from the outset, were only 300–400 yards distant, and in some cases only 100 yards. To close the distance still further, the Federals built approaches, zig-zag trenches, using sap rollers. There were thirteen such approaches, but the main ones hugged the roads. By June 2, the Federals had in places closed to within fifty yards. The bluecoats began mining operations, expedited by the fact that the loess soil allowed sappers to skip time-consuming shoring operations. The Confederates in turn countermined. A correspondent for the *Chicago Tribune* noted that a deserter "belonging to the Sappers and Miners says that the rebels have countermined some of our works. In some instances[,] the mines contain as much as one hundred pounds of powder in one place." Whether the deserter was a member of Wintter's company or simply part of a fatigue detail is not known.[43]

Henry Ginder was officially a draftsman, but during the siege he doubled as an engineer. He ran messages for Lockett and "went round among the citizens, and gathered up by one's and two's all the spades, shovels, pick axes, and grubbing hoes I could find, for we are very short of tools." Thousands of rounds of small arms ammunition were in a house, and there had already been close calls from artillery fire. On May 30, he was detailed twenty-five Black convicts from the city jail to begin digging earthen magazines near Stout's Bayou. Four had been completed by June 8, but by the 12th Ginder had been forced to condemn two of them on account of caving. "I have to act [as] engineer and overseer—have to go out with the negroes early before breakfast, and watch them and cuss them all day long; I mean scold them," he wrote his wife. Ginder became convinced that the fall of the city was "only a matter of time." He wrote his wife on June 16: "The enemy's mine is within 30 or 40 feet of our works; we are countermining & the workmen are now within six feet of each other."[44]

Lockett was aware of Union mining operations underneath the 3rd Louisiana Redan and began a countermine by sinking a vertical shaft, intending to work on an incline underneath the enemy's saps. The Federals could actually hear Confederate conversations and the sounds of picks at work. On June 25, the bluecoats struck first. An explosion erupted underneath the redan, creating a thirty-foot-wide, twelve-foot-deep crater and tearing off the vortex of the redan. The southerners, alerted to the mine, had prepared a secondary work and were ready. The 45th Illinois flooded in but could never make a penetration. The attempt cost 243 Federal casualties, versus 94 for the Confederates. The Federals began making a new mine from within the crater. Lockett began a countermine with a white supervisor and eight Blacks. He inspected the work at 2:30 on July 1 and found satisfactory progress. Thirty minutes later, Federal engineers ignited a massive explosion, obliterating the redan and killing or wounding over 230. In postwar years, a rebel recalled that the crater left a triangular shape, fifty feet across and twenty feet deep. Hesitant to repeat the incident of the 23rd, the Federals did not attack and contented themselves with a massive barrage. That afternoon Lieutenant P. J. Blessing was wounded.[45]

Throughout the night Captain Wintter, Lieutenant W. O. Flynn, and three men, all from the sapper and miner company, worked under a heavy fire to fill the breach, which was completed by daylight. The next day "we exploded one of our mines somewhat prematurely," Lockett wrote, "and we had ready for explosion 11 others, containing from 100 to 125 pounds of powder, and extending at a depth of 6 to 9 feet for a distance of from 18 to 20 feet in front of our works." The fuses were primed and everything ready, but on July 3 there was a flag of truce. The Federals likewise had preparations for mines at several points. By early July, the Federals had no less than ten approaches, ranging from five to a hundred yards from the rebel line. An all-out assault was planned for July 6, one that even Lockett admitted would have been successful, but Pemberton surrendered on the 4th.[46]

Criticisms

It could be argued that Lockett's defensive line kept Grant's army at bay for forty-seven days, but there were criticisms. Following the surrender, northern reporters had much to say about the defenses. One Chicago correspondent remarked that "everyone seemed disappointed" when the southern works were inspected, noting that, like the Confederacy itself, the works comprised "a bold front and weak reverses." Another reporter wrote that the angles and directions were all done well, and Lockett's ridge was higher than the opposing ground. Nonetheless, what appeared to be formidable forts turned out to be only "strong redoubts with a ditch a short distance to the rear." Nowhere were the defenses equivalent to the Union works, although the latter were prepared in haste. "Many places that looked quite impassable from our stand point two days ago," he concluded, "now seem to be only trenches two or three feet wide and a few feet deep."[47]

One Confederate writer heaped criticism upon southern engineering. Alexander St. Clair Abrams, an eighteen-year-old from New Orleans, served as a private in a Mississippi battery. He was discharged due to ill health, but could not make it out before the siege. He therefore took a position as a reporter for the *Vicksburg Whig*. Given his youth and lowly rank, his comments might be dismissed, but following his prisoner exchange his next newspaper, the *Atlanta Intelligencer*, published his highly critical pamphlet of the siege only months after the surrender. Given its wide circulation, the charges must therefore be examined.

First, Abrams wrote emphatically that between December 1862 and May 1863 little was done to strengthen the Vicksburg defenses. The slow progress was due to the fact that most thought that the line would not ever be used. He quoted "a prominent officer" who remarked that the rear works were "only thrown up to satisfy the public." It does in fact seem curious that, the limited tools notwithstanding, the works were far from complete after nine months, and those that were completed were badly maintained.

Second, there was a position to the left of the Jackson Road that should have been occupied by the Confederates. The Federals subsequently constructed Fort Logan on this hill, and they were able to enfilade the 3rd Louisiana Redan. "The position appears to have been entirely overlooked by our engineers, or its importance was very much undervalued," he wrote. Perhaps so, but even if the hill had been identified as a trouble spot, there was little that could have been done, given its distance from the main Confederate line. Third, Abrams leveled a general criticism of the line's weakness. It was "pronounced by the enemy's officers to be the most contemptable they had ever seen erected during the war." He added that "the supervisors of their construction could have known no more about erecting fortifications than we [readers] do; in fact, there was not one engineer in the army of Vicksburg who understood his profession thoroughly—they existed but in name." He claimed that after the surrender Union general James McPherson commented that "It was the rebels, and not their works, that kept us out of the city." In response, it can only be said that there were thirteen West Pointers and seven civil engineers in the rebel ranks. As at Fort Donelson, if mistakes were made, it was not due to a lack of professional talent.[48]

Captain Frederick Prime, the Federal chief engineer, also had his criticisms. "If at almost any point they had put ten or fifteen guns in position, instead of one or two to invite concentration of our fire, they might have seriously delayed our approaches," he concluded. He also added that the countermines of the southerners were "feeble ones, their charges always light." So ineffective were the Confederate field guns that the bluecoats rarely found it necessary to construct parapets of more than six to eight feet, rather than the standard fourteen. Prime also found Lockett's redoubts, with the exception of the Square Fort, "of weak profile." Despite Lockett's interlocking fields of fire and choke points, Solonick concluded that it was Yankee engineering ingenuity, specifically the molding of Mahanian theory and soldier improvisation, that won the day. In the end, Lockett's defenses were adequate only to a point. Even if Pemberton had stockpiled sufficient food supplies, his works would ultimately have been breached.[49]

6

ORGANIZING ENGINEER TROOPS

IN HIS BOOK *Engineering Victory,* Thomas F. Army Jr. concluded that the Union advantage in the Civil War came in the prewar educational system of the North that supported an industrialized economy and a labor system that rewarded ingenuity and mechanical skills. There was a resulting correlation between the manufacturing and educational systems in the North and what Army termed "remarkable engineering operations." The author concluded that the South "lacked competent engineers" and that the dearth of machinists, carpenters, masons, and mechanics "severely limited engineering operations."[1]

I would argue that the South *did* in fact possess—if not an abundant number—at least an adequate cadre of engineers and artisans. In 1860 Davidson County, Tennessee (Nashville), was home to ten civil engineers, four architects, three surveyors, six bridge builders, sixty-five machinists, fifty-one brick masons, fifty-nine stone masons, ninety-nine blacksmiths, and 447 carpenters, not to mention hundreds of laborers. There was sufficient talent in that one city alone to form an entire engineer battalion, *if* the Confederacy had been so inclined. Indeed, at both the Twin Rivers and the Siege of Vicksburg, it could be argued that the Confederates never lacked for engineer talent. When the 1st Michigan Engineers in the Army of the Cumberland was first organized, there were only four civil engineers in the ranks. The field officers and company commanders were bound together more by their ties to the Masons than their professions. One-third of the original recruits were farmers![2]

In table 6.1 I have counted the occupations of four Heartland states—Tennessee, Mississippi, Alabama, and Georgia. While there were those on the list who fought for the Union and those who served in other theaters, they

Table 6.1. Occupations of the Heartland, 1860

OCCUPATION	TENN.	MISS.	ALA.	GA.	TOTAL
Engineers (civil and mechanical)	374	226	446	421	1,467
Architects	25	5	17	11	58
Surveyors	32	29	29	21	111
Calkers	0	5	18	2	25
Boat builders	3	5	13	2	23
Blacksmiths	3,017	793	1,307	1,465	6,582
Bridge builders	15	17	3	4	39
Mechanics	596	943	1,797	1,465	4,801
Wheelwrights	976	336	2,386	592	4,290
Masons (brick and stone)	663	114	236	309	1,322
Carpenters	5,391	2,100	2,386	3,219	13,096

are offset by professionals and artisans from southern Kentucky and eastern Louisiana, Arkansas, and Virginia who served in the Heartland. Artisans such as machinists, foundrymen, molders, railroad men, shoemakers, saddlers, harness makers, iron workers, boiler makers, etc., were more likely be used in installations and foundries rather than as engineer troops and were not included.

The term "engineer" was loosely used in 1860, but even so there appears to have never been a debilitating shortage. Even if only a third of those counted were professionally trained civil engineers, that would still leave 489 available. This number then begs the question: If a pool of potentially hundreds of ante-bellum civil engineers were in the Heartland, why then was there seemingly such a dire shortage? The extent to which civil engineers served in other branches is not known, but clearly some did. Notable in this regard was Eugene F. Falconnet. Born in Switzerland in 1832, he came to America as a civil engineer in 1850, working with several smaller railroad companies. By 1861 he was in Nashville working for the Nashville & Northwestern Railroad Company. Unfortunately, he never used his technical talents during the war, serving first in Rutledge's Tennessee Battery and later as a major in the 14th Alabama Cavalry Battalion. Thirty-seven-year-old George W. Gordon attended the Western Institute in Nashville and graduated with a degree in civil engineering. He subsequently worked for the Nashville & Northwestern Railroad. Gordon opted out of engineering when the war began, becoming colonel of the 11th Tennessee and ultimately receiving a brigadier's commission. A number of civil engineers in the Heartland did not serve in the army, and indications are that they

Table 6.2. Civil Engineers Working for Heartland Railroads

Niles Meriwether	Mississippi & Tennessee Railroad
M. B. Pritchard	Memphis & Charleston Railroad
L. J. Fleming	Mobile & Ohio Railroad
George B. Fleece	Memphis, Clarksville & Louisville Railroad
R. C. Morris	Nashville & Chattanooga Railroad
Bentley D. Hassell	Memphis & Ohio Railroad
R. C. Green	Southern Railroad (Mississippi)

simply continued their civilian pursuits. Of the fourteen confirmed civil engineers in Memphis at the start of the war, four served in the army, one under contract to the army, and two continued as railroad civil engineers, but there is no record that the other seven either joined or contracted with the military.[3]

There were seventy-five confirmed engineers who worked in the Heartland (West Pointers with engineer backgrounds, civil engineers, and foreign-trained engineers) either as officers or under contract, not counting seven civil engineers who worked for railroad companies or surveyors and architects. The number of potential artisan engineer troops was over 30,000! The problem was the learning curve. Civilian engineers had no training in the construction of fortifications, laying pontoon bridges, drawing military maps, preparing abatis, lining off miles of trenches, and supervising both troops and slaves in large-scale projects.[4]

The 3rd Engineer Regiment

From the outset, the Confederates, and for that matter the North, were desperately short of formally trained military engineers. The Confederate Corps of Engineers (there was no separate topographical bureau) was established on March 7, 1861. There were two army organizations—the C.S. Regular Army, which was very small, and the much larger Provisional (temporary) Army. The Regulars counted only thirteen engineers, all of whom had formerly served in the US Corps of Engineers, but seven of those transferred to other branches. Of the remaining six, four had only limited engineer experience. By May, the Regular Army had increased to sixteen engineer officers and three engineer companies. Colonel Gilmer, as chief of the engineer bureau in Richmond, hoped to increase the bureau with "a few more proper men who can do something." Meanwhile, the dearth of qualified officers was filled by former civil engineers. "Many of these, though of clever attainments in their professions, had had no experience in military constructions up to the date of appointment," and they were forced to "reduce it to practice." There were by December 1862 ninety-three engineer captains in the Provisional Army.[5]

The problem was not just the shortage of officers, however, but the woeful lack of qualifications among those who had been given commissions. Of the ninety-three captains, according to Gilmer, twenty-four were "worthless men of broken down Virginia families—about twenty [commissions] to South Carolinians, no better than the aforementioned Virginians." He concluded that only a dozen officers were "good men—men who could lay claim to be called engineers." What Gilmer failed to concede was that most of the egregious engineering mistakes made, at least in the West, were by professionally trained West Pointers, not civil engineers.[6]

One reason why the Confederates were late in embracing the concept of engineer troops was that railroad repair continued to be relegated to private contractors, as it had been before the war. One of the primary bridge builders in the Heartland was the Anthony L. Maxwell & Son Company. Born in New York, Maxwell constructed bridges in Massachusetts before moving to Knoxville in 1850. By 1862, 100 hands were employed, with all of the tools and pile drivers necessary to complete large-scale jobs. Their most notable project was the rebuilding of the Nashville & Chattanooga Railroad bridge at Bridgeport, Alabama. Fortunately, the piers were still standing, but the Maxwell firm had to prepare and assemble the timber beams. The Tredegar foundry in Richmond was heavily employed in the manufacture of cannon, so the necessary iron parts (spikes, washers, connecting plate, rods) were cast by a foundry at Cartersville, Georgia. Marsh & Son later planked the bridge with over 80,000 feet of lumber for the passage of wagons and infantry.[7]

The Heartland counted four state-organized sapper and miner companies in 1861, but there were no formally organized engineer troops in the Confederacy, nor a single pontoon train. By December 1862, Gilmer was convinced that the southern armies could no longer rely on pioneer (fatigue) details, and that the Confederacy was in need of 4,000 formally organized engineer troops. These men, like their Federal counterparts, would repair roads, bridges, and fords, study topographical features, stake out trench lines, construct forts, perform reconnaissance missions, obstruct rivers, and lay pontoon bridges. They would be organized into companies (one to a division) of 101 rank-and-file, and the companies into regiments.[8]

The Army of Northern Virginia, as in the West, already had semi-engineer troops. By the summer of 1862 there was a company of Black pioneers that

Fig. 6.1. Destroyed Tennessee River Bridge. This 1864 photograph shows the destroyed Memphis & Charleston Railroad bridge at Bridgeport, Alabama. The Howe truss construction was performed by the Anthony L. Maxwell & Son Company of Knoxville in late 1862, but it was later destroyed by the Confederates when Bragg withdrew to Chattanooga in the summer of 1863. The pontoon under construction is by Federal engineers. Courtesy Library of Congress, LC-DIG-ppmsa-33480.

was proficient in bridge construction. Initially Robert E. Lee was receptive to the administration's idea of a formal engineer structure, but there were caveats. His engineer companies were then at half strength and operated only under the command of the division major general. When Congress formally mandated the size of each company and the regimental formation, Lee balked. The March 6, 1863, law would double the size of his engineer corps, making engineer regiments larger than some of his brigades. He further strongly objected to the companies being withdrawn from their assigned divisions and placed into regiments. The law directed that the men be veterans "chosen with a view of their mechanical skill and physical fitness," thus potentially eliminating many conscripts. In an attempt at appeasement, Secretary of War James Seddon replied that companies could be reduced to fifty men rather than a hundred, half of whom could be conscripts, and that they were to be in regimental formation only when not actively campaigning. He nonetheless would not yield on the regimental organization. If the outfits were retained as individual companies, they would "rapidly degenerate into mere drudges, scarcely better than camp followers, to be employed in menial service, burying the dead, etc." Such a result would cause the best officers to resign. General Orders No. 104, issued July 23, 1863, further refined the structure, including mandating at least two four-horse wagons for each company for camp equipment and entrenching tools, and a two-horse wagon for surveying instruments, stationery, maps, and drawing boards.[9]

In the summer of 1863, three engineer regiments and a battalion, the latter serving in the Trans-Mississippi, were organized. The 1st Engineers and four companies of the 2nd Engineers were assigned to the Army of Northern Virginia. The balance of the 2nd was scattered along the Atlantic and Gulf coasts. According to William Blackford, commanding the 1st Regiment, the men "were from 25–35, skilled in the use of tools in some way or the other, mechanics of all sorts. . . . miners, sailors, carpenters, blacksmiths, masons, and almost every other trade among us." The captains and field officers were all professional engineers, as were some of the lieutenants, and the noncommissioned officers "were well-educated men of special qualifications in some mechanical branch." Two companies operating in Lee's army were equipped as pontoniers. When the army was engaged in siege operations, the engineers served as sappers and miners.[10]

The nucleus of the 3rd Engineers was formerly Presstman's Engineer Battalion, organized in the spring of 1863. Two additional companies were organized in August. That summer a Texas brigade, which had been captured at Arkansas Post and exchanged, arrived at Tullahoma. Accompanying it was Captain A. W. Clarkson's company of sappers and miners, which became Company H. The regimental field staff comprised Lieutenant Colonel Presstman and Major John W. Green, commanding, Robert Percy, adjutant, E. P. Williams and Frank M. Duffy, assistant quartermasters, and J. L. Wesbrook and J. J. Goodwin, assistant surgeons. The regiment, which never had more than eight companies, was usually widely scattered. Companies A and D typically operated under Buckner in East Tennessee. Each company was led by a civil engineer. Captain Gloster, exchanged from his capture at Vicksburg, was tapped for Company C. Some officers (such as Mann and Smith) declined to command engineer troops, desiring to remain in the engineer corps. When Mann turned down Company G, Captain Cobb took his place.[11]

A few of the sixteen lieutenants (excluding Company E) were trained civil engineers, such as James J. Davies (Company G), an Augusta, Georgia, draftsman and engineer who had been on assigned duty on the Savannah River. William D. Prinz (Company D) was a former East Tennessee railroad civil engineer. Lieutenants T. S. Newcomb, Charles Foster, and George R. Margraves,

Table 6.3. Company Commanders, 3rd Engineer Regiment

COMPANY	COMMANDER	ASSIGNMENT
A	Capt. R. C. McCalla	Simon Buckner's Division
B	Capt. H. N. Pharr	Frank Cheatham's Division
C	Capt. Arthur M. Gloster	A. P. Stewart's Division
D	Capt. Edmund Winston	East Tennessee
E	Capt. William T. Hart	Southwestern Virginia
F	Capt. Waightsill A. Ramsey	Patrick Cleburne's Division
G	Capt. Robert L. Cobb	Thomas Hindman's Division
H	Capt. A. W. Clarkson	Sappers and Miners

all with engineering backgrounds, had long served as officers. Most of the lieutenants, however, were merely assigned "as engineers." Menefee Huston was listed in the 1860 Tennessee census as a "literary student." In 1861, Colonel Johnson noted that Huston would make "a good assistant," Thirty-three-year-old Lieutenant Thomas Jefferson Ridly (Company H) listed his occupation as a "machinist in the manufacture of cotton yarn." Lieutenant R. Riddely served only days before dying from unknown causes on June 30, 1863. Lieutenant P. W. Semmes (Company B), though not a professionally trained engineer, was a university graduate with top honors in mathematics and had a "liberal classical education." He had previously served in the 11th Louisiana.[12]

Each of the infantry divisions was assigned a company, to be comprised (as in the Federal engineers) of half artisans and half laborers. The two existing sapper companies in the Army of Tennessee (Pickett's and Winston's) were assimilated into the new organization. In July 1863, Polk directed that the mechanics in his two companies be comprised of two or three bridge builders, two or three blacksmiths, four wheelwrights, and forty-five carpenters. By September 9, four of the companies (B, C, F, and G) had still not received commissions for their officers, nor the proper equipment. Rives replied that the issues were being pressed, but in the meanwhile the companies should be treated "as engineer troops."[13]

Cobb's requisitions for Company G reveal the type of specialized equipment and tools needed: 4 four-horse wagons, 10 tent flies, 125 pounds of rope, 10 auger bits, 5 pounds chalk, 4 claw hammers, 4 oil stones, 1 bellows, 2 cut saws, 15 augers, 10 chalk lines, 1 anvil, 1 handsaw, 1 brace, 10 kettles, 2 hollow augers, 3 squares, 14 pounds steel, 3 planes, 1 set of chisels, 6 awls, 4 saw bits, 2 tape lines, 4 wedges, 1 wrench, 6 levels, 2 tongs, 6 axes, 2 rules, 100 horseshoes, 8

Table 6.4. Lieutenants, 3rd Engineer Regiment

COMPANY	LIEUTENANT
A	James S. Morrison
B	P. W. Semmes, Menefee Huston
C	T. S. Newcomb, Charles Foster
D	George R. Margraves, William D. Prinz
F	Henry Otey Minor
G	James B. Perkins, James J. Davies
H	John Palmer, R. Riddely, C. C. Gooden, R. H. Armstrong, T. J. Ridley

pounds horseshoe nails, 1 vice, 1 square, 10 buckets, 2 wall tents, 6 drawing knives, 1 try square, 5 sledge hammers, 2 gauges, 1 bench square, 1 tenet saw, 6 pad locks, and 1 buttress.[14]

The Road to Chickamauga

After six months of inaction, Rosecrans, on June 22, advanced on a wide front. He feinted against Bragg's left at Shelbyville, while the main army captured Hoover's and Liberty Gaps. Bragg aborted his last-minute counterattack by Polk's Corps and withdrew his army to Tullahoma. Polk ordered details from Cheatham's and Withers's divisions to report to Sayers and Morris and complete last-minute earthworks at Tullahoma, while Helm guided arriving units to their positions. On the 29th, Lieutenant Colonel Irvine C. Walker wrote his wife: "Our troops were industriously employed completing the entire line of fortifications." He felt that they "could easily have defeated the enemy had they attacked us." But Rosecrans had no intention to attack, at least not at Tullahoma. He instead struck with cavalry at Bragg's rear communications at Decherd. Coupled with the news that a column 10,000 strong was making a wide sweep on the right in the vicinity of Hillsboro, Bragg became rattled. Completely outmaneuvered and physically broken down, the army commander retreated to Chattanooga. Winston's sappers blew the Bridgeport railroad trestle in the face of the enemy and began obstructing the Tennessee River by sinking boats.[15]

Back in November 1862, Pickett's sapper and miner company had been ordered to build bateaux with the drawings made by Wampler, and to construct pontoon bridges at Battle Creek, twenty miles west of Chattanooga, and at Kelly's Ford. The company at that time mustered 105 present, with

Fig. 6.2. Captain Arthur W. Gloster. A former Tennessee civil engineer, Gloster saw service at both Vicksburg and in the Army of Tennessee. Courtesy Matthew Fleming, The Civil War Image Shop.

thirteen horses, thirty-two mules, and an ox. "The specialty of that company has been the making and care of pontoon bridges," Presstman noted. Polk's Corps crossed at the former on July 3, and Hardee's Corps at Kelly's on July 5. After some wrangling as to what to do with the boats, they were at last removed to Chattanooga.[16]

Hardee transferred to Mississippi in July. He was succeeded by Lieutenant General D. H. Hill, from the Department of North Carolina. Included in Hill's personal staff was his chief engineer, Captain Thaddeus C. Coleman, a twenty-six-year-old North Carolina civil engineer who had been attempting to obtain a transfer to the West. Due to Hill's subsequent poor performance and inability to support Bragg, he was replaced following Chickamauga, taking Coleman with him—at least for the time being. That summer, Captain Francois I. J. Thysseus, who previously served on Van Dorn's staff, was transferred to the Department of Tennessee headquarters at Chattanooga. It did not take long for his alcoholism to reemerge; this time charges were preferred. The captain subsequently requested and was granted a thirty-day leave of absence "for the benefit of his health." There is no indication that he reentered military service.[17]

Yankee mounted infantry caught the rebels napping and began shelling Chattanooga on August 21. The Confederate engineers had forty-seven pontoons lined up ready to be stretched across the 600-yard-wide Tennessee River. An attempt to salvage the boats was foiled by Federal sharpshooters on the riverbank. The Army of the Cumberland was on the march. Confederate leadership strongly suspected that Rosecrans planned to connect his left with Ambrose Burnside's 30,000-man army then approaching Knoxville. On August 28, Rosecrans did just the opposite, crossing the Tennessee River on Bragg's left. At Bridgeport, four companies of the 1st Michigan Engineers and troops of the pioneer brigade began constructing a 1,660-yard-long bridge—half trestle, half pontoon. Another pontoon bridge with 100 boats was laid at Caperton's Ferry. Amphibious landings were made at Shell Mound and Battle Creek.[18]

George Pickett's pontoon bridge, under construction at Kelly's Ferry, had to be hurriedly dismantled. David L. Kelly, one of the sappers, wrote to "Miss Honnell" on August 28, giving as his location "Building pontoon on Tenn. River." He noted: "[W]e had nearly completed [the bridge] but on Saturday night following we had to work till broad daylight Sunday moving the Bridge out of the River since which time we have moved out of town [Chattanooga] 2 miles."[19]

Pharr's Company B was sent scrambling. Between August 23–26 the men constructed a 206-foot-long bridge over West Chickamauga Creek. On September 8, the troops planked the 360-foot Oostanaula River railroad bridge at Resaca, Georgia. Between Catoosa Station and Ringgold, Georgia, the Western & Atlantic Railroad crossed Chickamauga Creek four times within two miles. On the 11th, Forrest's cavalry prematurely burned the bridges, creating a logistical nightmare. On the 16th, Captain Rowley and Lieutenant James D. Thomas were ordered to oversee the repair of the bridges, along with Companies B and C.[20]

Lieutenant Fergusson arrived in Chattanooga in late August. He was disgusted at the earthworks he saw, which he thought "were no possible benefit to anybody." A redan 300 yards from the railroad he dismissed as a "piece of folly." On September 6, he, Helm, and Thomas were dispatched to the Oostanaula River to hurry up the engineer troops working on the bridge. "All of the timber was cut and sections ready to be sunken, but not enough hands for rapid work," Fergusson wrote. He and Helm then crossed to the south bank and rode to Adairsville for a quick reconnaissance.[21]

Norquet accompanied Simon Buckner's corps to North Georgia that September. The Frenchman became involved in an infamous scene at McLemore's Cove on the 10th. Bragg planned a two-pronged attack against James Negley's exposed division at the foot of Lookout Mountain. The attack from the east stalled when Catlett's and Dug Gaps, obstructed by Confederate cavalry, could not be cleared in time, there being "no engineer company or working tools." Bragg still believed Thomas Hindman's division, reinforced with Buckner's Corps (two divisions), could accomplish the mission by attacking from the northeast. Based on misinformation, however, Hindman and Buckner expressed reluctance to follow Bragg's express order. Major General Thomas Hindman sent Norquet to the army commander to relay his fears. The major gave an incoherent report in broken English. An angered Bragg sent Norquet back to Hindman with an order to attack. The major garbled his message, relating that Hindman should execute his own plan and Bragg would support him. Fearing that they were in a vulnerable situation (in reality three Confederate divisions to one Federal), Buckner and Hindman spent the afternoon looking for a way to retreat. Buckner's engineer company (Richard McCalla's) was eventually sent to clear Catlett's Gap. At 1:00 p.m. Presstman arrived from army headquarters, inquiring if Catlett's Gap had been cleared in the event of a retreat. This added to the speculation that Bragg was having second thoughts. The attack never developed and Negley's bluecoats escaped.[22]

The massive Battle of Chickamauga was fought September 19–20. Having received reinforcements from Mississippi and Lieutenant General James Longstreet's Corps of the Army of Northern Virginia, Bragg won the victory, but the staggering losses (18,454), the fact that Rosecrans's army escaped to Chattanooga, and the subsequent Confederate high command in-fighting mitigated the strategic importance of the victory. On the 24th, Polk's chief of staff received a brief note through the lines: "I was taken prisoner on Saturday evening [19th] about 3 o'clock. We are treated very well, everyone civil and kind to us." The message was signed by Edward B. Sayers. A Union brigade commander, Colonel William Grose, noted that "he [Sayers] surrendered to me in person." Captain McCalla was at Chickamauga River during both days of the battle—"we worked all Saturday night making forts and Bridges and all day Sunday fortifying," all the while listening "to the roar of cannon and musketry." Sayers never returned to duty. He lived out his life in St. Louis, dying in 1881 at the age of forty-nine. Three weeks after the battle, Presstman asked Lieutenant Fergusson if he would ride out to the battlefield and make a special survey of the defensive works of Rosecrans's final stand. He went to the site "and found only a few piles of rails for about 200–300 yards and I reported showing map of location."[23]

Buckner's Corps returned to East Tennessee, but indications are that Norquet still fumed from the McLemore's Cove incident. Bragg admitted that he had spoken harshly to him. Whether what happened next occurred for that reason, or for the major's seeming demotions and frequent transfers, or whether he simply lost the stomach for the war is not known. What is known is that he took a large sum of Confederate money meant for East Tennessee bridge construction, exchanged it for Tennessee bank notes, and deserted to the enemy wearing civilian clothes. In postwar years, Norquet lived outside of Chicago, serving as an architect for the Illinois Central Railroad. He died in Evanston, Illinois, in 1890, at the age of sixty-five.[24]

In early October, the men of Company B, 3rd Engineers, worked on the fortifications around Chickamauga Station. In mid-month, Captain Rowley oversaw operations at the Oostanaula River railroad bridge, where Companies B and F planked the trestle for the passage of wagons. For nearly three weeks that month, Companies A and D labored to construct three small bridges over Chickamauga Creek. "It has just been an ugly job as the men had to work in mud and water underneath and the rain from above," Captain McCalla wrote his wife.[25]

In early November 1863, Brigadier General Danville Leadbetter assumed command as chief engineer of the Army of Tennessee. As a West Point graduate, class of 1836, he spent sixteen years in the Regular Army, much of it in the engineer corps. He was transferred to Mobile in 1852, and while there met and married a wealthy widow. He subsequently resigned his commission to oversee her properties. Prior to his assignment to the Army of Tennessee, he served in Mobile and the Department of East Tennessee. Brigadier General E. P. Alexander, one of Longstreet's officers and himself a former member of the US engineer corps, was not impressed. "Leadbetter was one of the old captains of my old corps, the U. S. Engineers. . . . By education, experience, & position (he was now a general of engineers) he ought to have been as good a military field engineer as there was in either army." An officer who knew Leadbetter in the Old Army recalled that he was known as "the skinflint of the army."[26]

The East Tennessee Campaign

In late August 1863, Major General Ambrose Burnside's Army of the Ohio invaded East Tennessee, occupying Knoxville by September 1. Throughout the autumn Confederate forces remained thinly scattered; the Federals took advantage. Brigadier General James M. Shakelford's cavalry division struck toward Bristol, Tennessee, on October 15, destroying the East Tennessee & Virginia railroad bridges over the Holston River at Zollicoffer and at the Watauga River bridge, advancing all the way to Bristol, where three engines and twenty-eight boxcars were torched. The Anthony L. Maxwell & Son Company had

Fig. 6.3. Danville Leadbetter. Following the war, Leadbetter fled first to Mexico and then to Canada, possibly to avoid a war crimes trial for hanging several of the East Tennessee bridge burners. He died in 1866 in Canada, and his body was returned to his adopted state of Alabama. *Miller, The Photographic History of the Civil War,* vol. 5, 257.

by now removed their valuable tools and pile-driving equipment to Georgia. Replacing the bridges would thus be a job exclusively for the engineers. The project did not get off the ground until early December, when Captain J. M. Robinson was dispatched as the on-site engineer.[27]

McCalla's Company A completed the Zollicoffer bridge by January 12, 1864, and then began on the Watauga project ten miles downriver at Carter's Depot. "We will close this bridge [Watauga] in three or four days," McCalla wrote his wife on February 4. "In a week from today we shall have the cars [running] to Morristown and considering that every bridge on the [rail] road with only one exception was destroyed by the enemy or our own men. I think it has been rebuilt in an astonishingly short time and the Engr. troops deserve some credit at least."[28]

Meanwhile, in late October and early November Bragg dispatched Lieutenant General Longstreet's Corps, which had been operating with the Army of Tennessee, to retake Knoxville. He began removing his troops from Lookout Mountain on the night of November 4. While there may have been valid strategic reasons for the move, Bragg's clear intent was to rid himself of the eastern corps commander, a member of the anti-Bragg clique and who surreptitiously sought army command for himself. Bragg would hold his position on Missionary Ridge, while Longstreet would quickly drive Burnside out of Knoxville and then return to Chattanooga. But what if the East Tennessee Campaign stretched out indefinitely? The question was best answered by Earl Hess: Bragg was "unwilling to take the offensive" and "seemed willing to do nothing more than fritter away his resources in secondary objectives."[29]

From the outset, the campaign was fraught with problems, not the least of which dealt with the engineers. Longstreet requested a topographical map of the region, but he never received one. Buckner sent "some inaccurate maps of the country, along the Holston [River]—all that he had." In postwar years, Longstreet admitted that he received a topographical outline of the country between the Hiwassee and Tennessee Rivers, which was in the rear of his field of operations. He added that "Buckner was good enough to send me a plat of the roads and streams between Loudon and Knoxville."[30]

Longstreet had a good engineer on his staff, Major John J. Clarke, but he of course knew nothing of the East Tennessee terrain. Thirty-one-year-old Clarke, a former Petersburg, Virginia, railroad civil engineer, would survive the war, only to be killed in 1880 in a freak accident in a small New Jersey town near New York City, when he was struck by a runaway train. E. P. Alexander commanded Longstreet's artillery, but he had previously served as an engineer in the Old Army. For the time being, he would have to serve in dual roles. Alexander liked Clarke, whom he described as "a great friend" and "a fine clever fellow." Longstreet requested, and subsequently received, the assistance of Brigadier General Leadbetter, who had been in his new role as Bragg's chief engineer only a short while. The request nonetheless made sense, Leadbetter having previously served in Knoxville. The brigadier did not arrive until November 25. Another engineer serving in Longstreet's Corps was Lieutenant Thomas J. Moncure, a thirty-year-old graduate of the Virginia Military Institute, who had been a railroad civil engineer prior to the war.[31]

The East Tennessee engineer companies were busily engaged. McCalla's and Winston's outfits (Companies A and D, 3rd Engineers), were at Charleston, Tennessee, on October 30 constructing a pontoon bridge and rebuilding the railroad bridge over the Hiwassee River. "We will finish the trestle bridge in four or five more days, and where we will go then I do not know, but should not be surprised if we go to Loudon," McCalla wrote his wife. The companies were still at Charleston on October 9, guarding the bridges and preparing timbers for the destroyed Tennessee River bridge at Loudon.[32]

With the Loudon bridge wrecked, Longstreet's advance came to a halt. Theoretically, of course, a pontoon bridge could be built almost anywhere the terrain permitted, *if* there were wagons to haul them, which there were not. The pontoons would have to be offloaded from railroad flatcars as close to the river as possible. Alexander and Clarke spent two days reconnoitering the vicinity for the best location. The engineers finally decided on Hough's Ferry, not far below the destroyed bridge. "The bend was high & concave on our side of the stream, & the railroad track within a few hundred yards," wrote Alexander, later adding that the overland trail was through wooded terrain. Getting the reserve pontoons, which were then being built by Pharr's Company B in Dalton, Georgia, was a task in itself. Several trains of flatcars were required. By late October, Presstman was in Atlanta with four of his engineer companies preparing the prefabricated parts for a permanent bridge. The pontoon bridge was laid on the night of November 14. "Clarke worked very hard & successfully, &, by soon after sunrise in the morning, the troops were able to pass, & the artillery & trains followed," Alexander recalled. A staff officer wrote that the pontoon bridge was "a sight to remember." "The current was strong, the anchorage insufficient, the boats and indeed the entire outfit quite primitive, and when lashed together finally to both banks it might be imagined a bridge; but a huge letter 'S' in effect it was its graceful reverse curves."[33]

The campaign came to a climax with the Siege of Knoxville and the November 29 attack on Fort Sanders. Longstreet committed 6,000 men in the dawn assault. In the darkness, the rebels encountered wire entanglements, causing them to fall. "They rose up to stagger on a few more paces, and then go down again," wrote a Federal. The attempt to scale the slippery fort parapet, eighteen feet in height and at a seventy-degree angle, proved disastrous. The southerners fell back in disarray, having lost 813 casualties to the Federals' fifty.[34]

The question arose as to Leadbetter's role in the attack. Longstreet appears to have placed great stock in the brigadier's advice—it would be a mistake. Alexander did not mince words—"I have never been so disgusted in my life," he wrote, noting that Leadbetter had "no appreciation of ground," "appeared to become demoralized" on November 27, and "after not finding the place he hunted up above & I cannot recall his making another suggestion or saying another word that day." As far as the November 29 attack, Alexander remarked that "I will go to my grave believing that Leadbetter devised it & imposed it upon Longstreet, & he afterward preferred to accept the responsibility rather than plead that he had let himself be so taken in." So much for Leadbetter.[35]

Fig. 6.4. Fort Sanders. Originally started by the Confederates and named Fort Buckner, Fort Sanders was greatly strengthened and renamed by the Federals. After the failed Confederate assault on November 29, 1863, mounds of rebel dead and wounded lay in the moat fronting the fort. Courtesy Library of Congress, LC-DIG-ppmsa-33483.

Retreat to Dalton

Following the southern victory at Chickamauga, Bragg unsuccessfully attempted to besiege the Federals in Chattanooga. Grant replaced Rosecrans, numbers of Federal reinforcements arrived, and Lookout Mountain was recaptured. Now on the defensive, Bragg established a defensive line at the base and along the summit of Missionary Ridge. The slight earthworks comprised logs, rocks, and earth, hastily thrown up due to a lack of tools and the rugged terrain. "This new line was run out by an engineer officer, whose name I have forgotten. He appeared to be in a great hurry, and said that he had much to do," recalled Brigadier General Arthur M. Manigault. Not liking his line, the brigadier suggested that he would oversee the defense of his sector. "This he [the engineer] agreed to do, and seemed much pleased at being relieved of some portion of his work, but at the same time told me that his instructions were to run the line on the highest point or outline of the hill." The line thus stretched along the geographic crest, rather than the military crest (the most direct line of fire), giving a great disadvantage. On November 25, the Federals assaulted in force, the rebel line was shattered, and a rout ensued.[36]

Bragg's defeated army retreated south to Dalton, Georgia. Several wagons were lost due to the many holes and deep ruts in the road. A correspondent complained loudly that no engineers or pioneers were to be seen. He wondered when the authorities would see the necessity "of organizing an efficient corps of engineers, including pioneers, bridge and boat [pontoon] builders, etc." Unfortunately, all of Company A, sixteen men of Company D, and thirty-seven men from Company F, all on duty in Charleston, Tennessee, when the army withdrew, were blocked from reuniting with the Army of Tennessee. Company G blew the bridge over Chattanooga Creek on the road leading to Chickamauga Station. The next day the men of D again engaged in demolition, this time firing the bridge over Chickamauga Creek on the road leading to Ringgold. That December Walter J. Morris, the army's most experienced topographer, was relieved of duty for four months due to ill health. By the end of 1863, the engineer troops at Dalton totaled 435 officers and men.[37]

Command changes occurred in early 1864. General Joseph E. Johnston replaced Bragg as army commander, and Lieutenant General John Bell Hood, having recovered from his severe Chickamauga wound, took over corps command from Breckinridge, who had been dismissed by Bragg. Thaddeus Coleman, whose requested transfer to the West had at long last come through, became Hood's chief engineer, with Lieutenant Helm as assistant engineer.[38]

Once at Dalton, the 3rd Engineers went to work on several projects. Company G constructed a 103-foot-long, 8-foot-high bridge over the third crossing of Mill Creek on the main Dalton Road. The army's two pioneer (as opposed to engineer) companies, led by Captain John Oliver and a Captain Gilip, engaged in cutting saplings to corduroy 2,300 yards of the Dalton-Resaca Road. At that time only Oliver's pioneers and Company D, 3rd Engineers, were at work on the project. Fourteen miles from Dalton, Gilip's company was cutting a new road. To completely repair the highway, Major William Pickett, Hardee's inspector general but also a civil engineer, recommended a detail of 400 men with three engineer officers. The Sugar Valley Road to the west of the main highway was in even worse condition. "I do not think a large army train can possibly get through on this route without a great loss of wagons," Pickett concluded. Hardee submitted the report to Leadbetter with the endorsement: "Major Pickett is an engineer by profession and very accurate in his statements." A detail of 500 men and Clarkson's Company H, 3rd Engineers, were sent to complete the repairs.[39]

Work commenced in mid-March on two dams on Mill Creek, leading directly through Mill Creek Gap. The purpose was to flood the gap in the event of a large-scale Federal advance. The task was assigned to Oliver's pioneer company. The needed timber was cut and sawed at Tilton, Georgia, nine miles south of Dalton. The pioneers used the Western & Atlantic railroad bed, which was twenty to thirty feet higher than the lowest parts of the gap, as the dam. They then plugged the stone culverts (one of which survives to this day) and built a flume, as a means to raise and lower the water level. The project required over five weeks.[40]

Meanwhile, Companies C, D, and G were ordered to Atlanta to construct pontoon boats. The train was designed to bridge the Tennessee River in the event Johnston went on the offensive. Some 135 wagons (Leadbetter had earlier estimated 150 wagons) were required, each drawn by four mules, to haul the bateaux, chess (planking), rope, anchors, and tools. The specific type of pontoon is not known, but in August 1863 John W. Glenn, "a young officer of great energy and skill" at the Enterprise, Mississippi, engineer office, was charged with making pontoons "according to the plan adopted by Mahan." He was transferred to Dalton in April 1864 to establish a shop for the manufacture and repair of tools, and it is possible that he assisted in the construction of the pontoons in Atlanta. Rives informed Captain Grant on January 18, 1864: "Presstman will present to you the drawings of pontoons and pontoon wagons [flatbeds]. Assist him in every way. Iron parts can be manufactured in Atlanta shops." A member of Tucker's Mississippi brigade was "creditably informed" that some wagons were to double as pontoons. "[E]ach body and frame is so constructed as to be used for pontoon bridges whenever it may be necessary to use them," wrote Arch McClarin.[41]

Johnston radically reduced the army's baggage train. The engineer's headquarters was assigned one wagon, but each of the companies was reduced from four wagons to one. Each brigade was allowed a tool wagon, with approximately 125 tools, as compared to an average Federal brigade wagon with 350 tools. Work on the Western & Atlantic Railroad defenses had been going on since the fall of 1863, under the direction of Lieutenant B. F. Roberts, a Confederate engineer assigned to the Resaca defenses, Captain Calvin Fay, the former Memphis architect, and twenty-six-year-old Captain T. G. Raven, an English-born graduate of a military school who had come to America for business pursuits. He contracted pneumonia in late 1863 and died.[42]

Presstman returned from a leave of absence in January. There were new arrivals that spring, including forty-year-old Captain George H. Hazlehurst, assigned to the army staff. As a civil engineer, he worked extensively on railroads in the Heartland, including the New Orleans, Jackson & Great Northern, the Southern Railroad, and the Nashville & Chattanooga. Three lieutenants also arrived in May—twenty-six-year-old George R. McRee, D. W. Currie of North Carolina, and G. H. Browne. Twenty-two-year-old Thomas E. Marble, a graduate of the Western Military Institute of Nashville, excelled in math while in school and was granted a commission in the engineers. Ordered from Virginia to the Army of Tennessee, Marble was killed before he arrived in Georgia.[43]

7

THE MAPMAKERS

THE US TOPOGRAPHICAL ENGINEERS did not formally come into existence until the War of 1812. The concept was disbanded in June 1815, but reinstated a year later. The formal organization of the US Corps of Topographical Engineers was authorized in 1838. It comprised the top students by ranking from West Point. The primary peacetime mission was the surveying and mapping of rivers (including vital maps of the Mississippi River later used by the Confederates), roads, railroads, harbors, and coasts (the last leading to the formation of the US Coast Survey). Antagonisms and jealousies frequently existed between the engineers and topographers, the last perceiving themselves as the proverbial stepchild. During the Mexican War the topographers played a significant role, but between that time and the Civil War the bureau received little recognition and governmental financing and fell into obscurity. The work during that period was largely in exploring the western frontier, making boundary surveys, and laying out the route of the Union Pacific Railroad.[1]

The Civil War brought a demand for reliable maps, and remarkably few were to be found. In Tennessee, Samuel A. Mitchell's 1859 state map was available, but it showed little detail beyond major highways and railroads, and there were no topographical features. County maps and civil district maps proved scarce. An old Jackson Purchase map revealed details about western Kentucky, but duplicates had to be traced—a laborious task. As late as the summer of 1862, Albert H. Campbell, chief topographer for the Army of Northern Virginia, wrote: "Incomplete tracings or fragments of the old 'Nine Sheet' Map of Virginia were probably all that our commander [Lee] had for guidance."[2]

Some primitive maps were drawn early in the war. Nothing is known of cartographer Albert Martin, but sometime probably in the summer or early fall of 1861 (Fort Pillow had not yet received that name) he drew a map, not to scale, of the Mississippi River fortifications from Memphis to Ashport, Tennessee. Perhaps the most important geographical information imported were the huge swamp areas, what today would be categorized as flood plains, from Fort Harris to Island No. 36 and from Fort Randolph to Fort Cleburne (Fort Pillow).[3]

Many engineers drew tracings and rough map sketches, but by 1862 a few more sophisticated maps were being prepared. The Fremaux maps of the roads from Corinth to Pittsburg Landing and the Battle of Shiloh have been previously mentioned. Fremaux is known to have made maps of northern Mississippi, but none survive. There is an additional 1862 map by an unknown Confederate cartographer, approximately 19.5 x 23 inches, which reveals the fortifications around Corinth, camp roads, cultivated fields, railroads, the Danville Road, and Hardee's headquarters. As Polk prepared his after action report for the Battle of Stones River, Walter J. Morris put the finishing touches on his watercolor map of the battlefield. He had previously surveyed the area and merely added in troop dispositions, acquiring names of Federal units from prisoners. The map that accompanied Bragg's report has also survived, although its authorship is unknown. Like the Morris map, it is drawn to a scale of one inch per 2,000 feet. Although both maps were sufficiently correct to give an overview of the terrain and troop positions, neither approached the sophistication of Federal cartographers.[4]

Little is known of H. A. Pattison beyond the fact that he was a civil engineer (he signed his name with C. E.) and that he prepared maps in Pennsylvania years earlier. His February 1863 map of the Vicksburg defenses (1:26,200) shows relief drawn by hachures and is one of the more professional-looking Confederate maps. Following the Battle of Chickamauga, Morris drew at least three maps of the battle with the notation: "Based upon notes taken by me upon the Battle Field to the east of the State Road during the engagement and added to and corrected after the battle and revised and additional information obtained from the official reports of the officers of both armies." The maps (1:2,217,600) are somewhat difficult to read, due to the fact that units appear more than once with different time notations. By comparison, the Chickamauga maps of twenty-nine-year-old Captain Edward Ruger, former civil engineer of Janestown, Wisconsin, and topographical engineer with the Army of the Cumberland, are more encompassing and refined. Ruger noted that he based his maps in part on a rebel battlefield map.[5]

One of the army's primary mapmakers was twenty-four-year-old, German-born Conrad Meister. Arguably his best work was his "Map of Perryville and Surrounding Country by the Reconnaissance of Engineer Corps." Meister operated out of Knoxville at the time, and acquired his information from map sketches made by others. The composite map encompassed the area from Louisville to Lexington and south of Perryville. On a scale of one inch to three miles, it included all principal roads, creeks, towns, and rivers. Meister made two maps of Stones River, one penned on December 29, 1862, and the other a campaign map that included the sector from Nashville to Triune, east to Murfreesboro, and from Murfreesboro generally north to Lebanon. In late February, Meister was dispatched to Huntsville to map the area, and between May 23 and June 8, 1863, he drew a map of Shelbyville, Tennessee, and the fortifications, although it has not survived.[6]

On July 25, 1863, a rebel deserter crossed the Union lines at Stevenson, Alabama, and was taken to Philip Sheridan's headquarters. The prisoner turned out to be Conrad Meister. He openly revealed sensitive information, including the fact that the pontoon bridge at Kelly's Ford had been taken to Chattanooga, and he offered a sketch of rebel battery positions around the city. He also brought maps. "The officer has a very accurate sketch of the country between Bridgeport and Chattanooga, extending back some considerable distance," Sheridan reported. "He has valuable maps and information of Chattanooga and the country south and west of there."[7]

A reporter of the *New York Herald* happened to be in Winchester, Tennessee, on August 3 and conversed with Meister. He submitted the following article for his readers:

BRAGG'S TOPOGRAPHICAL ENGINEER DESERTS HIM

I saw the individual today at headquarters of this department, now located at this place [Winchester]. He is a young German, whose family and friends live in Brooklyn, and who was impressed into the rebel service at Memphis a couple of years ago. It is hardly to be denied that the service had a particular charm for him, though he would have much preferred that of the Union. With much skill of the pencil and fond of sketching the adventuresome life in the army and had its attractions for him. To desert was to denounce that life, and it is not to be doubted he left Bragg with some feeling of regret.

He states that Bragg's topographical corps is very large and excellently organized. All of the topographers are in camp at Bragg's headquarters, and are detached with commands on the march whenever information upon their line of march is desirable. This system is found to work most admirably, the topographers not being interfered with Division commanders who may wish work done for themselves. The following are Bragg's chief engineers:

Captain S. W. Steele, chief engineer
Lieutenant H. C. Force, assistant chief engineer
Lieutenant H. [A.] H. Buchanan, assistant chief engineer
Lieutenant J. K. P. McFall, assistant chief engineer

For obvious reasons the name of the deserter in this case is not given. He is an exceedingly intelligent gentleman, and since he has been at headquarters has finished a map of Chattanooga from actual field notes, giving all information of importance to us. The heights of the hills, prominent peaks, relative altitude of the bluffs on either side of the river, location of batteries and camps, are given with what he says is great accuracy.

He brings with him a large map of the country from Bridgeport to Chattanooga—a map of the greatest importance to us. It was while engaged in the duty of mapping this district that he crossed the river [Tennessee] and came over to our side. He was dressed in his best suit of gray, and his portfolio, or haversack, was filled with maps, notes, and caricatures of different rebel officers—Polk, Hardee, etc. He had also some sketches of the Stones river battlefield. He is to be allowed to be returned to his home in Brooklyn in a few days.[8]

Gilmer, on July 22, reminded Presstman that "Maps of the late campaign [Stones River] need to be made but we have no draftsman in Richmond to send you. Detail qualified soldiers in the ranks." E. G. Anstey, a former Memphis musician detailed from the 15th Tennessee, had already been drawing maps for Major General Benjamin F. Cheatham. One tracing from a Federal map was that of Nashville and the surrounding counties, drawn at a scale of 3.5 miles per inch. Anstey's map of North Georgia, the original of which survives at the Chickamauga-Chattanooga National Battlefield Park, represents northeast Georgia, from Chattanooga to Kingston, Georgia, and extends slightly into northwestern Alabama. The road that ran diagonally between Calhoun and Rome, and the highway east and west of Rome, show individual residences by name.[9]

Despite being a civil engineer, Captain John S. Tyner, commanding a company in the 1st Confederate Cavalry, was not transferred to the engineers until 1863. Born in Macon, Georgia, he later moved to Harrison, Tennessee. He drew several maps for General Joseph Wheeler, some of which survive. Another one of Wheeler's engineers, Silvanus W. Steele, also drew some maps, at least one of which survives. Twenty-seven-year-old Joseph H. Haney, who hailed from Van Buren, Arkansas, initially served in the 3rd Arkansas, but after Shiloh he transferred to the Louisiana Washington Artillery (5th Company). Being a former civil engineer, he was detached after Stones River to draw a map of Confederate battery positions around Tullahoma, Tennessee. He also drew a map of troop dispositions at Hoover's Gap and Liberty Gap. Assistant engineer John S. Stewart made maps for Wheeler's cavalry (several survive in the Alabama Archives), but no further information is known.[10]

On November 1, 1863, Captain Wilbur Foster was directed to oversee a massive map project that would extend from Missionary Ridge to Atlanta and Rome, Georgia, for a width of ten miles on either side of the Western & Atlantic Railroad. Fortunately, the state of Georgia was covered by prewar government surveys and section lines and stakes, which were usually easily located. Skeleton sketches of the state were also available to the capital at Milledgeville, which, though limited in scope, proved helpful. A total of ten assistants would work in the field, most of them with civil engineering backgrounds: Valentine Herman, Felix R. R. Smith, Napoleon B. Winchester, J. W. McGuire, Andrew Buchanan, Frank Gaines, John F. Steele, Henry C. Force, William W. Fergusson, James D. Thomas, James H. Allen, and draftsmen J. Louis Tucker and Charles Foster. Fergusson recalled the list as being Steele, Force, Foster, Helm, Pickett, Gaines, Buchanan, McFall, and Winchester for field work and Anstey and Tucker as draftsmen. Steele and John C. Wrenshall were reassigned after a short time, and Pickett was shot in the heel, leaving him disabled.

Each of the officers was assigned either a section or part of a section of the map and, equipped with a pocket compass and protractor, was directed to draw in roads, bridges, fords, houses, villages, and bridle paths. When completed, the sectional maps were added to the composite map and sent to the chief engineer in Atlanta, Lemuel P. Grant. The officers worked without an escort and were always subject to bushwhacking or capture. Indeed, one of the assistants was captured and another had to ride for his life to escape. Yet another engineer was detained by a group of six zealous locals. Major General Henry D. Clayton had a number of maps of sections of Georgia that survived the war; whether or not these are tracings from the master copy is not known.[11]

By the beginning of the Atlanta Campaign, Johnston had a continuous map that included every road, residence by name, creek, river, cleared land, and basic topography for forty miles wide on either side of the railroad. Lieutenant Buchanan, an avid Johnston supporter, told of how the army commander could ride a large section of a battlefield without a compass and detect at a glance any errors in a map. Indeed, one engineer was dismissed for errors in a map of an important district that he had ridden over twice.[12]

Although none of his maps survive, Charles Foster of Marion County, Tennessee, was known to be a cartographer. Born in London in 1802, he immigrated to the United States in 1835. The family lived in several places in the East and Midwest before settling in Cincinnati in the 1840s. His diary (1835–1853) included forty color pen and ink plates, some of which survive. Foster became an illustrator for several agricultural journals and prepared some almanacs. Sometime after 1845, he moved to Tennessee and became a land speculator and farmer, while continuing his art work. During the war he worked primarily out of the Knoxville office. He continued his drawings after the war, concentrating on scenes in Marion County, where he died in 1889.[13]

Thirty-two-year-old Lieutenant William W. Fergusson was a schoolteacher from Carthage, Tennessee. He joined the 2nd Tennessee as a musician (he could play the violin and piano) and then his outfit was transferred to Virginia. There his "peculiar taste and extraordinary qualifications" as a topographical engineer were noticed. In the spring of 1862, he was reassigned to Corinth, but a fever and jaundice, something which would hamper him throughout the war, kept him in the hospital. He served in Lockett's office as a draftsman during the Vicksburg Campaign.

A scene occurred following the Battle of Chickamauga that underscored the difficulties in reconnoitering and platting maps. Bragg summoned Fergusson, Force, and Steele to his headquarters to draw maps of Lookout Mountain and to make recommendations for defenses. He would compare their sketches to an old map in his possession. "The fact is Gen'l Bragg had little comprehension of the points of the compass from Point Lookout and yet, when attempting to explain from the map furnished him, to explain or to point out particular points," even those within viewing distance of his headquarters, "it was next to an impossibility to comprehend fully what he desired to accomplish or the direction of the work to be done," groused Fergusson. It was not the first time the army commander's poor map skills had been called into question. Brigadier General Edward P. Alexander noted that it was whispered that he could not understand a map and that "it was a spectacle to see him wrestle with one, with one finger painfully holding down his own position."[14]

Each of the officers was to make his own map and come back with recommendations. Steele, being the oldest, recommended only the defense of Lookout Point, and his plat showed only 150 yards of trenches. Force's map included the Summertown Road and Lookout Point, but nothing south of the road. "This is not what I want," the general snapped, and then turned to Fergusson and asked to see his work. The young lieutenant awkwardly presented his map. Bragg carefully studied it and compared it to the original map in his possession. He asked how many yards south of the Summertown Road would have to de defended? Fergusson answered: "One thousand yards." "How wide is the mountain where you have located the works?" Bragg inquired. "About 1,880 yards," Fergusson answered. "Is it possible? I had no idea the mountain widens and so quickly as that—could you not obtain a line that would reduce the length by bringing it a short distance nearer?" he queried. Fergusson answered that he could, but that would lose the highest point on the mountain, which would surely be occupied by the enemy, who would then fire down into Confederate lines. "We can't afford that," the general concluded, and he ordered that the trenches be commenced.[15]

Far less is known of J. Louis Tucker, who apparently held the unlikely occupation of teamster prior to the war. As a member of an outfit called the Warsaw Rifles, he was captured in Missouri in 1862. After his exchange he was assigned to the engineer office in Tullahoma, where he drew several maps. Fergusson wrote that he was twenty-six years old, dark, with a fat face. He claimed to have been born in Virginia, but some doubted it. He was educated in New York, loved the ladies, and "the Yankeeisms in his manner and pronunciation" could be detected. Captain Sayer is known to have drawn some maps, but only one has survived—a May 6, 1863, Shelbyville area ink map on oilcloth folded to pocket size.[16]

Charles F. Baker, who had been serving in the cadet battalion of the Georgia Military Institute in Marietta, Georgia, was detached and assigned to Captain Wrenshall's map production office in Macon in August 1864. Baker was responsible for plotting, drawing, and duplicating maps of notes made by W. J. McCullough of Marietta and his assistants on reconnaissance missions. One of Baker's maps, showing the district around Jacksonville, Alabama, is shown in *Confederate Veteran Magazine*.[17]

Initially the Army of the Cumberland's engineer office and cartographers were hardly better prepared than the Confederates. Under the leadership of Captain William E. Merrill, the reorganized topographical corps began using civilians, many of whom had no particular qualifications. He initially found "the headquarters office almost destitute of assistants or means of doing works, and the engineers of the different commands utterly ignorant of what they were wanted for, and equally unsupplied with means of doing anything." Merrill completely reorganized the corps, and by the summer of 1863 it far outpaced his southern counterparts. He increased the strength to fifty officers by July, with assignments to brigades, divisions, and corps. Each was equipped with a prismatic compass and portable drafting kit, and supplied with information maps with sketches of the most available terrain features. Brigade topographers, with all the collected maps and reconnaissance notes, reported every Monday to the division topographers, and they in turn reported each Wednesday to the corps topographers, who reported to Merrill.[18]

Even the Army of Northern Virginia seems to have been far advanced from their sister army in the West in map making. Five months before the Foster map project began in North Georgia, Lee ordered a similar task for northern Virginia. Two or three survey parties were formed with the necessary instruments and, starting from Richmond, radiated out toward the front lines, each taking a sector. Work was often interrupted to go to other locales. The project concluded in the spring of 1863. Eventually thirteen parties were engaged, and by the close of the war the Virginia counties around the Rappahannock to Wilmington, North Carolina, were mapped.[19]

It was Albert Campbell who came up with the idea of photographic reproduction of maps. Initially photographers advised him that it was impractical, but it was successfully tried. Tracings of maps were made in India ink and copied via sun-printing in frames made for that purpose. Placing the negatives onto glass was a complicated process requiring a number of chemicals. The sections were pasted together to form the larger map. The end result, though crude by modern standards, worked well enough.[20]

Photographic reproduction of maps for the Army of Tennessee was done by thirty-six-year-old Andrew Jackson Riddle, a Columbus, Georgia, photographer best known for his prints of the Andersonville prison camp. Born in Baltimore, he moved to Georgia sometime in the early 1850s and opened a Daguerreian studio. He was arrested three times by Federal troops, twice while attempting to smuggle photographic supplies through the lines. By August 1864, Riddle had photographically reproduced sixty maps 17.5 x 18 and 8 x 10. On September 18, 1864, Captain Buchanan wrote a note to Lieutenant Glenn. "Col. Presstman directs me to say to you he wants one dozen more copies of the map which was photographed some time since: the one embracing the country between the Atlanta and West Point R. R. and the Chattahoochee River as far as Newman. Please send them to Hd. Qrs. as soon as they can be procured." Glenn's endorsement read: "Capt. Wrenshall will please have the proper negative sent to the photographer." Photographic reproduction was also done in Selma by assistant topographical engineers J. F. Knight and Lee Mallory.[21]

The overwhelming majority of Confederate maps were hand-drawn, and of varying degrees of quality. The Federals delved into map making far earlier than the southerners, they were unquestionably better organized, and the end product many times revealed a higher degree of talent. Nonetheless, the Confederates got the job done. It is regrettable that less than two dozen original Heartland maps survive today.

8

WE WANT ENGINEERS

FROM DECEMBER 1863 to April 1864 the Army of Tennessee stockpiled supplies and returned to the fundamentals of training, marksmanship, and discipline. Mass religious revivals, overwhelming reenlistment numbers, and Johnston's perceived military genius all resulted in soaring morale. Numbers remained an issue, with only 54,000 in and around Dalton, but reinforcements were on the way—Polk's Army of Mississippi (16,800), a brigade from Savannah (2,337), and two regiments of Georgia State Line troops (550), a total of nearly 74,000, with a scattering of regiments and a division of the Georgia militia yet to come. An infantryman informed his homefolks that "a new spirit seems to have been infused into the army since Genl. Jo Johnston took command."[1]

As always, several command changes occurred in the engineers. Captains Henry C. Force and John F. Steele transferred to the Army of Northern Virginia and Savannah, respectively. Forty-year-old Captain George H. Hazlehurst, a former civil engineer who had worked for several southern railroads, arrived on Hardee's staff. Johnston thus began the campaign with nineteen professionally trained military and civil engineers in the immediate Dalton area.[2]

The war was entering a new phase that would test the limits of the modest engineer corps. Both sides would henceforth engage in the construction of battlefield fortifications that stretched for miles. Trenches, redoubts, and abatis were certainly nothing new; indeed, Mahan's *Complete Treatise on Field Fortification,* published in 1836, had even been reprinted by the Confederates in 1862. "Semi-permanent works," as Mahan referred to them, had been used in the Western Theater and to a degree shaped tactics, but the sheer scale in the Atlanta Campaign would eclipse anything imaginable. The days of Shiloh-,

Table 8.1. Engineer Corps, Army of Tennessee, May 1864

OFFICER	ASSIGNMENT
Brig. Gen. Danville Leadbetter	Army of Tennessee, chief engineer
Capt. J. K. P. McFall	Army of Tennessee, assistant engineer
Maj. George B. Pickett	Hardee's Corps, chief engineer
Capt. George M. Helm	Hardee's Corps, assistant engineer
Capt. George H. Hazlehurst	Hardee's Corps, assistant engineer
Capt. Thaddeus Coleman	Hood's Corps, chief engineer
Capt. Silvanus W. Steele	Wheeler's Corps, chief engineer
Lt. J. S. Tyner	Wheeler's Corps, topographer
ENGINEER TROOPS (FIELD STAFF)	
Maj. Stephen W. Presstman	3rd Engineer, commanding
Maj. John W. Green	3rd Engineer
ATLANTA OFFICE	
Capt. Lemuel P. Grant	Bureau chief
Lt. John W. Glenn	Engineer Supply Depot
COLUMBUS, GA. OFFICE	
Capt. Theodore Moreno	Bureau chief
J. C. Wrenshall	Topographical office

Stones River-, and Chickamauga-style battles involving open field maneuvers were passing. Defensive warfare now became dominant. Johnston openly declared his strategy to be "on the defensive," and to attack only under favorable circumstances. While he, with consultation, selected defensive lines, the actual design and lining off of the works reverted back to the engineers.[3]

The Atlanta Campaign

A Federal reconnaissance in late April and another in early May 1864 led many to suspect that the long-anticipated Union advance on Dalton was at hand. The bluecoats had made little movement by May 5, but they were observed cutting roads, making causeways, and building bridges. Although Johnston made no countermoves, he kept his army on full alert. On the morning of the

7th, a Kentuckian jotted in his diary: "I think I shall now have something more stirring to put in my journal than church goings and keeping a meteorological table." That morning found Oliver's pioneer company on a ridge clearing brush for the artillery when heavy fire suddenly erupted in the distance. Orders came for the pioneers to hurriedly proceed to the Taylor Creek bridges and destroy them as soon as Wheeler's cavalry passed. They arrived to find the troopers crossing in a continuous stream. Both bridges were torched, and the men retired for the evening. At 11 p.m., however, they were again called out, this time to cut trees and obstruct Mill Creek Gap. The Atlanta Campaign had commenced.[4]

Sherman's army group included George Thomas's Army of the Cumberland, John Schofield's Army of the Ohio (actually a corps), and James McPherson's Army of the Tennessee. It is difficult to make a direct comparison of the opposing engineer structures. At its peak, the Army of Tennessee's engineer battalion comprised six companies in North Georgia (B, C, D, F, G, and H) and two pioneer companies, while each of Polk's two infantry divisions in the Army of Mississippi, on their way from Demopolis, Alabama, had an engineer company, and James Cantey's division coming from Mobile had a pioneer company—for a total of eight engineer and three pioneer companies, perhaps 625 or so men. Almost all of Sherman's divisions had a pioneer company, and the Army of the Ohio counted an engineer battalion. The Army of the Tennessee was assigned the 1st Missouri Engineers, although the regiment did not report until the end of the campaign. There were two pontoon trains, the Army of the Cumberland having the 58th Indiana with 600 feet of bridging and the Army of the Tennessee fielding a pioneer unit with 800 feet of bridging. The 1st Michigan Engineers remained in the rear repairing bridges and roads and building blockhouses.[5]

Sherman outflanked Johnston at Dalton, and the action quickly shifted south. Presstman, on May 10, was dispatched to Resaca to mark out a defensive line that extended from the Conassauga River, across the Western & Atlantic Railroad and Dalton highway, then curved south along Camp Creek, to the Oostanaula River. Polk's Corps opportunely arrived and began digging trenches, but the fortifications were hastily prepared, primarily due to the lack of entrenching tools. There also appeared to be "no engineer officers handy to tell the men what to do," concluded Hess. The battle raged May 13–15, when the Federals crossed the Oostanaula River at Lay's Ferry, west of Resaca. If successful in flanking the rebels, Sherman could strike the railroad in the Confederate rear; Johnston had no choice but to withdraw. The army crossed at midnight on the 15th, Hardee's and Polk's Corps (Army of Mississippi) by the railroad bridge, and Hood's Corps by the pontoon bridge. Despite the hurried nature of the Resaca breastworks, Orlando Poe, Sherman's top engineer, was impressed. "I don't think I exaggerate in saying that they were five miles long," he wrote his wife, "and in many places were composed of two and even three lines." It was nonetheless obvious to Poe that such near impregnable positions could always be flanked, as had now happened at Dalton and Resaca. Within four days the Federals had rebuilt the railroad bridge.[6]

There were no large mountain ranges south of the Oostanaula, so a stand at Calhoun became impossible. Based on a map prepared by the engineers, however, Johnston determined to take a position a mile or two north of Adairsville in Oothcaloga Valley. There, or so he believed, the valley presented a sufficiently narrow bottleneck that his smaller army could deploy with the heights protecting his flanks. A quick on-sight inspection nonetheless revealed the valley to be far broader than portrayed, so Johnston withdrew. Interestingly, the incident was never mentioned in his after action report, only in his memoirs. The map in question has not survived, so it is difficult to know precisely what was before him. It is known that an engineer detail, comprised of Thomas, Force, Fergusson, Buchanan, Herman, Gaines, Winchester, McFall, and Foster, made a reconnaissance at Adairsville on December 9, 1863, and at that time prepared a map, almost certainly the one in question. It is difficult to see how the cadre of professionals could have so badly miscalculated the distances, if in fact that was the case.

The 1877 J. T. Dodge-H. H. Ruger map reveals the width of the valley to be only two miles. The one-mile-long Confederate works were bisected by Oothcaloga Creek, and open on both flanks, but not by much—a quarter-mile on the right and one mile on the left. The two roads leading into the valley roughly formed a reentrant angle, with the earthworks at the apex. A Johnston staff officer noted in his diary on May 17: "Enemy reported turning our [left] flank beyond Oothcaloga Creek." A makeshift line could easily have been extended up the Bowen Church Road, with artillery crowning the hills around the Jenkins's farm. Such a deployment would have made Johnston's line no more than two and a half miles in length. Given that he had nine infantry divisions by that time (Samuel G. French's division was still one day out, but Red Jackson's cavalry division arrived on May 17), his statement that "The breadth of the valley here exceeded so much the front of our army properly formed for battle, that we could obtain no advantage of ground" simply does not stand up under scrutiny. Indeed, when Johnston took a position at the Chattahoochee River in July, his line extended six miles, more than double the width of Oothcaloga Valley. Confederate William Trask described the Adairsville deployment thus:

Table 8.2. Engineer Corps, Army of Mississippi

OFFICER	ASSIGNMENT
Capt. Walter J. Morris	Army of Mississippi, chief engineer
Lt. Gottfried G. Gutherz	Army of Mississippi, topographer
Capt. John Baptist Vinet	Loring's Division, chief engineer
Capt. Deitrich Wintter	Army of Mississippi headquarters (as of early July 1864)
Capt. ? Wilkerson	Army of Mississippi, assistant engineer (July 1864)

ENGINEER TROOPS	
Capt. W. A. C. Jones	Company engineers, commanding
Capt. J. A. Porter	Company engineers, commanding
?	Pioneer Company, Cantey's Division

"The whole army was drawn up in a narrow rolling valley with high hills on either flank and I thought our position an admirable one." Johnston, for whatever reason, simply did not want to fight at Adairsville, and the work of the engineers has been vindicated.[7]

South of Adairsville, Johnston planned a trap at Cassville. In a lengthy conversation with Lieutenant Buchanan, who had previously surveyed the area, he was assured that the topography was "open, and unusually favorable for attack." The projected May 19 assault was aborted, however, when Hood's flank accidentally collided with two Yankee cavalry divisions. The army withdrew to a defensive position east of the Cassville Road, with Hardee's Corps on the left, astride the railroad, Polk's corps-sized Army of Mississippi in the center, and Hood's Corps on the right, the last two atop a 140-foot-high ridge, with a fifty-foot-deep gap separating Polk's right and Hood's left. Half of Matthew Ector's brigade occupied the gap, with Hoskins's Mississippi Battery fifty yards in advance. A concern arose as to whether Hoskins's guns would be enfiladed by a 10-pounder Parrott battery (I, 1st Michigan), more than a mile to the north. Johnston believed that traverses could be constructed for flank protection and the position made secure. Polk directed his chief engineer, Walter Morris, to investigate. Morris, having returned to duty following a four-month sick leave, literally stepped off the train at Cassville at 4:00 p.m. His subsequent report indicated that traverses would not protect the position from enfilade fire. The opposing ridge was on a line 23 degrees southwest, forming an angle with Polk's line of about 25 degrees. He concluded that the

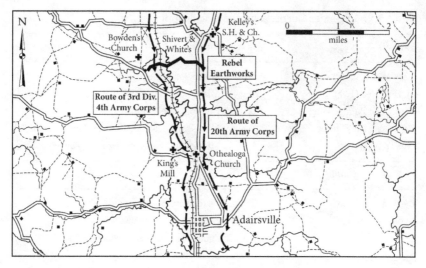

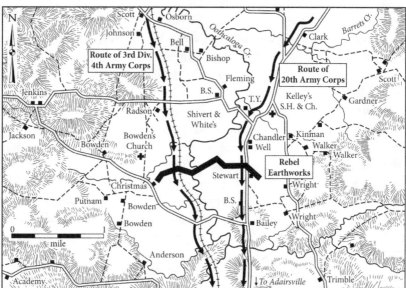

Maps 4a and 4b. The Adairsville Affair. In his memoirs, General Johnston accused the engineers of misrepresenting the width of the Oothcalooga Valley, three miles north of Adairsville, Georgia, thus foiling his plans for a stand. The 1877 J. T. Dodge-H. H. Ruger North Georgia Map No. 2 reveals the truth. The valley was no more than two miles in width (as evident in Map 4b, *below*), with commanding hills on either side. Johnston's postwar criticism of the engineers was not valid. Courtesy Library of Congress Geography and Map Division, G3921.s85.US.

enemy guns could not only reach the ridge, but also 600 yards beyond it. In consultation with Polk and Hood, neither of whom believed the position could be held, and unwilling to take the offensive, Johnston reluctantly withdrew that night across the Etowah River.[8]

Hess discovered an error in Morris's postwar map, as presented in Hood's memoir. "Comparing it with modern topographical maps, it is clear that Morris drew a greater angle in the ridgeline, bent at the gap, than was warranted," although Hess did not believe that this was a deliberate misrepresentation. "Because of the fact that the ridgeline south of the gap lay a bit farther west than the position north of the gap, any Confederate line would have been at least partially exposed to enfilade fire, but probably not as much as Morris believed based on his somewhat faulty survey of the terrain."[9]

On May 22, a New York infantryman ventured out to inspect the rebel works at Cassville. "They were the finest we had seen up to that time and it must have taken much labor to build them," wrote Rice C. Bull. "Johnston's army could not have reached there more than a few hours before we were in front of them so they must have been constructed prior to the start of the campaign. They were no ordinary breastworks that could be built overnight but strong elaborate fortifications with redoubts and abatis in front." Bull's comment is puzzling since there is no indication that the Confederate works had been constructed sometime earlier. There are occasional references to "temporary redoubts," but the indication from this veteran is that these works were much more substantial. If they had indeed been built earlier, this would shed skepticism on Johnston's often repeated Cassville theory of a last-minute trap. Until additional evidence is forthcoming, the incident remains unresolved.[10]

The troops snaked southward through Cartersville, crossing the Etowah River on the two pontoon bridges and the railroad and country road bridges. The engineers then pulled up the pontoons and burned the bridges, although a disgusted correspondent noted that only a portion of the trestle was destroyed and, with the piers remaining, the railroad bridge could be easily replaced in a short time. According to a Johnston staff officer, however, a foul-up occurred; the railroad bridge was never supposed to be destroyed. When Sherman did not immediately pursue, Johnston momentarily considered recrossing the Etowah. "Our pontoons have been ordered back, and the duplicate railroad bridge over the Etowah, all ready to be put up, has been sent forward for construction," noted a correspondent. Assembly of the bridge began on May 21 and continued the next day, but the army commander lost his nerve and ordered the duplicate destroyed on the 24th.[11]

Johnston withdrew five miles south of the river to the Allatoona Heights, hoping that Sherman would follow the railroad and assault him there. The Union commander instead crossed downriver and maneuvered toward Dallas, about fifteen miles southwest of the railroad, and where the terrain was thickly wooded and visibility at time restricted to 100 yards. "Early Saturday morning [May 21], the line having been indicated by the engineers, our men busied themselves erecting rifle pits," wrote a Tennessean in Hood's Corps. No sooner was the work complete when they were told to start marching once again. Supplied with one of Walter Morris's hand-drawn maps, drawn to a scale of three miles to an inch, Johnston shifted Hardee's and Polk's Corps toward Dallas. By May 25, the Army of the Cumberland approached the Confederate position at New Hope Church. The subsequent Union assault was predictably shattered. An attempt to turn the rebel right at Pickett's Mill on the 27th likewise ended in Union disaster. With his lines stretched thin, Sherman now returned to the railroad.[12]

Beginning in late May and early June 1864, both sides began to prepare temporary breastworks every time they stopped. These were impromptu affairs done by the men themselves and usually without tools. Within ten minutes a breastwork could be erected of logs and limbs, mixed in with rocks and stones. Simply by using the material at hand, a barrier "seemed to grow out of the ground," noted Brigadier General Arthur Manigault. Although the engineers and pioneers did engage in digging trenches, they typically were confined to other duties. Relating the backbreaking work of Oliver's pioneers, Hiram Williams chronicled the events the last week of May. On the 23rd and 24th the men repaired a mile and a half of the Cassville-Allatoona Road. On the 25th they constructed arbors for the wounded sustained in the Battle of New Hope Church, and the next day they buried the dead. The company was ordered to Alpheus Baker's brigade of Stewart's division on the 27th to work on rifle pits. "Horrid to work on breastworks during an engagement with the enemies lines within 150 yards," Williams jotted in his diary. The following day the men cut a road through the woods for ambulances and artillery, a task not completed until May 30. The final day of the month, twenty men of the company cut a road on Lost Mountain.[13]

The Kennesaw Line

In early June, Johnston deployed in an extended position that stretched from Lost Mountain on the right to Brush Mountain on the left, with a position on Pine Mountain, about 1.25 miles in advance of the center. Major General

Samuel G. French of Polk's Army of Mississippi, whose division initially held Lost Mountain, had been in a bit of a mood for some time. "We want engineers," he frustratingly wrote in his diary. There was also the issue of tools; he needed them and none were available. He required from Morris a survey of the lines for the Army of Mississippi. On the 7th, French was directed to take his engineer company and division tool wagon and report to either Presstman or Morris on the Burnt Hickory and Marietta Road at the foot of Kennesaw Mountain. The next day, June 8, Morris directed Captain Vinet to turn in all of his division tools, but the wagon was parked at corps headquarters. "I cannot understand the object of keeping these tools idle. I cannot construct my line without tools," French complained. The next day he received an order to take a new position to the right of Lost Mountain; he was not happy. He wrote in his diary: "This morning Maj. Presstman explained the ground for my line. It is a weak, faulty, miserable line. The engineer took all my tools yesterday, so today I am unable to construct any works."[14]

For the next week the rain proved unrelenting. On the morning of June 14, the sun finally broke through. A group of high-level officers, including Johnston, Hardee, and Polk disregarded warnings that Federal artillery was zeroed in, and the group climbed atop Pine Mountain for an inspection. A Federal battery, seeing the cluster of officers, fired a couple of rounds, one of them striking Polk and nearly cutting him in half. He was temporarily replaced by W. W. Loring, and the Army of Mississippi merged into the Army of Tennessee as a third corps. The next morning Johnston abandoned Pine Mountain. Hardee, holding the left, was forced to reposition his corps on the 16th. This meant that Loring's and Hood's line ran west to east, while Hardee's line extended north to south and faced west. This created an angle of the type which Lockett had carefully avoided at Vicksburg and which the Federals now quickly exploited. Hardee's trenches on the right (W. H. T. Walker's division) and Loring's left division (French's) could be and subsequently were enfiladed. Johnston ordered Presstman to immediately prepare another line south of the Mud Creek-Brush Mountain line, but until that was done the rebels had no choice but to tough it out.[15]

Presstman's new and much stronger line, forming a huge semicircle west of Marietta, was occupied by the southerners on the night of June 18–19. Loring's Corps, on the left, occupied the Kennesaw Ridge (Big Kennesaw at a 691-foot elevation, Little Kennesaw at a 400-foot elevation, and Pigeon Hill measuring a 200-foot elevation), Hardee's Corps held the center, from the Burnt Hickory Road to the Powder Springs-Marietta Road, and Hood's occupied the right, on either side of Olley's Creek. The weak link in an otherwise near impregnable position was a salient near John Ward Creek that would

Fig. 8.1. Engineer Camp of the Army of Mississippi. Drawn by Lieutenant Gottfried "Fred" Guntherz of the engineers, the drawing depicts the engineer camp at Kennesaw Mountain on June 22, 1864. Courtesy Alabama Department of Archives and History, Q26194.

Fig. 8.2. At work in camp. Based on a self-portrait of Lieutenant Gottfried "Fred" Guntherz, the engineer in the drawing is probably Guntherz himself. Courtesy Alabama Department of Archives and History, Q26192.

forever become known as Cheatham's Hill. Captain Hazlehurst admitted as much to his brother-in-law, the colonel of the 66th Georgia. The engineers had been forced to extend the line in that sector to an inverted V-shape in order to prevent the Federals from occupying a hill. "I hope you won't get into one of them [salients] as there the main fighting is apt to be," Hazlehurst noted. Unfortunately, in the darkness, the engineers had placed the defensive line on the geographic crest, rather than the military crest, thus creating a blind spot for rebel infantrymen. Cheatham's Tennesseans worked throughout the night to strengthen the position, constructing a parapet and placing abatis thirty-yards deep. The subsequent Yankee assault on the morning of June 27 was easily repulsed, with the Federals losing 3,000 to the Confederates' 700.[16]

Sherman, who had vowed repeatedly never to assault a fortified position, had done so; he would never attempt it again. He now reverted to his flanking maneuvers, forcing Johnston to once again fall back. northern correspondents and officers took the opportunity to examine the now abandoned works at Kennesaw. The trenches were "cut down square and true, and the parapets shaped with the square and plummet." The works, built "in the most approved style of engineering," evoked "the admiration of all military men with whom we talked." An Ohio soldier described to the homefolks how the trenches were riveted with heavy timbers that were covered by an embankment of earth from four to sixteen feet thick. Beyond the abatis were trees felled with the tops outward, "so as to make a tangled and almost impassable surface." Anyone who examined the works, he concluded, could not but help but "be amazed at the achievements of this [southern] army."[17]

The day after the battle, John Green wrote his mother from engineer headquarters near Marietta. "As yet my command [3rd Engineers] has not been called upon to face the fire in battle. At present it is very much scattered." He further related that some of the engineer companies never made it back from East Tennessee, and that several companies were at the Chattahoochee River, in charge of the pontoon train at that place. This left him but three companies in the field.[18]

Chattahoochee River Defense Line

With the abandonment of the Kennesaw Line virtually certain, the engineers were sent scrambling. On July 1, Oliver's pioneers of Stewart's division constructed some "forts," but no sooner did the work near completion than orders were given to once again fall back. "So the hard work on the forts as usual was done for nought, as not a gun has been in them, much less fired out of them,"

Hiram Williams disgustedly wrote. That night the company marched four miles and "stoped at a creek [Nickajack] and built a large signal fire, where the troops could see to cross the stream."[19]

On the night of July 2, the Confederates withdrew to the six-mile-long Smyrna Line, with the right on Rottenwood Creek and the left near Nickajack Creek, where the line refused south for a mile and a half. Brigadier General Manigault was not pleased, believing the line to be weak, irregular, and with many salient points. In the midst of the foul, rainy weather, the troops worked to piece together makeshift defenses, with skirmishing and artillery fire continued. At least a portion of the line was more substantial, however, for Sherman's chief engineer labeled it a "well-built" position with connecting salient and gun embrasures, and a Confederate colonel wrote his wife that "there is very little danger for we now have excellent works." Meanwhile, Captain Wintter and his engineer assistant, a Captain Wilkerson, and a draftsman, were sent south to examine the roads and ferries and "all information you can."[20]

The Smyrna Line was always designed to be a holding action, and on July 5 Johnston withdrew once more to the Chattahoochee River Line. Interestingly, the most studied defensive position in the campaign was actually not designed by an engineer, but by Brigadier General Francis A. "Frank" Shoup, Johnston's chief of artillery, who the army commander had frequently used on engineer duty. Initially Shoup claimed that he devised a novel concept meant for a single division—"a huge tete-de-pont," as he termed it, to cover the railroad bridge on the north side of the Chattahoochee River. A system of twelve-foot-high arrowhead-shaped log redans, resembling "a three-sided log cabin minus a roof," according to Michael Shaffer, each meant for eighty riflemen, would be spread about 60–170 yards apart and connected by an eight-foot-high stockade barricade. About halfway between the "shoupades," as they became known, was a reentrant angle where there was a two-gun redan, thus creating a crossfire. The shoupades and redans were mutually supportive and offered enfilade and interlocking fire.

When the brigadier explained the concept to Johnston, however, he mentioned a four-mile-long-line, from just above the Western & Atlantic Railroad bridge, "to give full space within for his whole army." Was the Chattahoochee River Line thus meant for a division-sized bridgehead, keeping the balance of the army mobile, or was it meant for the entire army? While Shoup never clarified this obvious contradiction, the subsequent construction clearly showed that he meant the entire four-mile-long line as the tete-de-pont. Historian Craig Symonds may well have been right when he claimed that the idea was to leave a single corps on the north bank, while the other two corps would be free to pounce on Sherman as he crossed the river. Unfortunately, Shoup, at least in his postwar article, never specifically declared such. Two weeks into the project, Johnston sent orders to extend the line another three miles (actually only two miles, for a total of six miles) to the left to cover the Mayson-Turner Ferry crossing. Shoup expressed frustration, stating that the army commander clearly did not understand the concept. Yet Shoup's own explanation was blurred and ill-defined.

Shoup was assisted by Major Wilbur Foster, who had been in Georgia on various assigned missions since January. A thousand Blacks labored two weeks on the project. The left of the line was anchored at Nickajack Creek with a seven-gun redoubt, and on the right at the Chattahoochee River with an eight-gun redoubt. On the south side of the bridge, across the river, were three batteries of irregular-shaped redoubts. The soldiers never grasped the unconventional concept, and the shoupades became the butt of jokes. Sherman's chief engineer later dismissed them. The overall line, however, he found to be well laid out and the strongest system yet encountered. After four days, Johnston abandoned the position. In retrospect, the weaknesses were obvious. The rebels would be fighting with their backs to a river, which would spell disaster in the event of a major breach. Given that the entire army was needed to occupy the trenches, the line could be easily flanked. The project was oddly experimental for such a critical time. In the end, it proved an enormous waste of time and labor that could have been better spent strengthening the Atlanta defenses.[21]

Captain Winston, described as a "clever gentleman and an accomplished officer," along with his engineer company and with the assistance of Captain Coleman, Hood's chief engineer, had built fortifications and laid two pontoon bridges back in the spring. As Johnston's army crossed on the night of July 9–10, the engineers, assisted by a detail of seventy-five infantry, removed the lower pontoon bridge. With the army safely across, the two permanent structures and the upper pontoon bridge were all destroyed.[22]

Push the Work Night and Day

Following the rebel evacuation of the Chattahoochee River Line, Sherman paused for a week. John N. Witherspoon, assistant quartermaster of the 3rd Engineers, took the opportunity to write his friend Captain Robert P. Rowley. He apologized for his slow response, but confided: "Presstman is out on a 'rampage.'" He later wrote a more light-hearted missive. "Nor is it necessary as were all are on the alert to view once more that [Rowley's] huge mustache which has slain so many of the fair sex and filled so many beardless youth with envy." On a more serious note, he wrote that all men were thrilled at the captain's recent promotion to major, which unfortunately came with a transfer to the Trans-Mississippi.[23]

Dissatisfied with Johnston's constant retreats, Davis fired him and placed Hood in command. Following through with Johnston's plan to attack when Sherman crossed Peachtree Creek, Hood assaulted on July 20 as the Yankees crossed. Although possessing superiority in numbers (26,000 versus 20,000), the poorly coordinated attack failed. Realizing that he would be withdrawing into the Atlanta defenses, Hood directed Presstman to make an immediate inspection. The chief engineer reported the next morning, July 21, that the northern city defenses were only partially completed, that they had been built on low ground, were too close to the city, and were altogether "totally inadequate." One must wonder, of course, why engineers had not previously raised red flags. Presstman was ordered to immediately stake off a new line that could be occupied that night. The result was that L. P. Grant's roughly circular works around the city were squared on the northwest and northeast sectors. This, of course, created dangerous right angles, subject to enfilade fire. Presstman was not teaching a geometry class, however; he was preparing a defensive line and he had less than ten hours to do it. The colonel of the 43rd Mississippi would declare the position "a badly constructed line of trenches."[24]

Advised on the afternoon of July 21 that Sherman's left flank east of Atlanta was "in the air," Hood determined to attack. While A. P. Stewart's (Polk's/Loring's old) Corps and Cheatham's (Hood's old) Corps withdrew into the city defenses, Hardee's Corps conducted an exhausting night march to strike Sherman's flank. The overly ambitious plan resulted in the Battle of Atlanta on the 22nd, in which the Confederates were repulsed with gruesome losses. An attempt by Stephen D. Lee's (Hood's/Cheatham's old) Corps to defeat Sherman's flanking maneuver three miles east of Atlanta resulted in the Battle of Ezra Church on July 28 and another lopsided Union victory. Throughout early August Sherman shelled the city, an act which Poe openly opposed to Sherman.[25]

As the fighting dragged wearily on, several command turnovers occurred. Leadbetter, who had been in bad health for some time, was transferred to Mobile. Shortly thereafter, however, he went on furlough and tendered his resignation on August 20. He was replaced by Martin L. Smith on August 1. Smith had been chief engineer in the Army of Northern Virginia, but Lee, apparently on Davis's request, reluctantly released him. Assisting him would be thirty-nine-year-old Lieutenant Colonel Bushrod W. Frobel. A West Point graduate who acquired his engineering skills at Annapolis, he served aboard

Fig. 8.3. Captain Robert P. Rowley. "All are on the alert to view once more that huge mustache," a fellow engineer jokingly wrote the captain. *Confederate Veteran,* April 1899.

a ship in the US Revenue Service. At the start of the war he briefly served in the Virginia state navy, such as it was, later transferring to the artillery. Frobel came to the attention of General Chase Whiting, who placed him on his staff. Later transferred back to the artillery, he became Hood's division artillery chief, being promoted to major in July 1862 and lieutenant colonel in November, when he was sent to Wilmington. Ordered to the West by the War Department, he arrived in Atlanta in early August 1864, where Hood "received me cordially," according to Frobel. Also reporting in August was Lieutenant Miles M. Farrow, a graduate of the Hillsboro Military Academy in South Carolina, where he excelled in Mahan's military and civil engineering and surveying. He relieved Captain Rowley of command at West Point, Georgia, where defense works were under construction. Lieutenant D. W. Currie of North Carolina, about whom little is known, also reported that summer and was assigned to Lee's Corps.[26]

The day after his arrival, Frobel, along with M. L. Smith and Presstman, rode out to the eastern portion of Atlanta to examine a vulnerable part of the line. The three dismounted and cautiously crept outside the trenches, where the sector was particularly vulnerable and subject to enfilade fire. After crawling about for an hour, Smith and Presstman returned to the city, but Frobel rode to a large redoubt on the Georgia Railroad. He was advised (wisely) to dismount, as sharpshooters not 100 yards off were taking shots. He examined an embrasure lunette with parapets and sandbags mounting six to eight guns. Before departing, he observed that the trenches were half-filled with a foul greenish water.[27]

Other turnovers occurred. Walter Morris transferred back to the Department of Mississippi, Alabama, and East Louisiana, being replaced by Major Foster. Two experienced engineers fell captive—Felix R. R. Smith on April 13 and James Thomas on July 19, and Major Green was shot in his middle thigh on July 22. In late July and early August, Captain Hazlehurst was on assignment gathering railroad iron. Following the fall of Atlanta, he was transferred to the Department of Mississippi, Alabama, and East Louisiana. Captain Joshua A. Porter, a former railroad civil engineer commanding an engineer company in A. P. Stewart's Corps, was assigned to corps headquarters.[28]

As the fighting shifted south of Atlanta, the engineers exchanged their shovels for muskets. On August 13, the battalion, estimated at 600 strong (in reality only 495), was ordered in the redoubts at East Point. "Red" Jackson's cavalry division covered the army's flank in that sector, but an infantry regiment was needed for support in the event they were driven back. Yankee horsemen 4,700 strong struck Jonesboro on August 19, ripping up about two miles of track. The next day Hood dispatched Presstman with a brigade to repair communications. Within two days the trains were once again running. Back in Atlanta, Frobel noted on the 21st: "All night long the sound of shovel and pick might have been heard in the trenches." The next day Hood personally inspected the work, giving orders to "push the work night and day until completed." He returned on the 23rd and "personally urged that all diligence should be used." An entire brigade was detailed as an additional workforce.[29]

As the bluecoats continued to shell Atlanta throughout late August, Frobel worked against rain and snipers to strengthen the defenses. While inspecting a portion of the lines with Brigadier General George Maney on the 21st, a 20-pounder shell exploded dangerously close to the brigadier, showering Maney with thick mud and water. "Hang those dirty villains," he exclaimed, "They have nearly spoiled my *Sunday Clothes!*" Five days later Frobel glanced into the bleak no-man's land and recorded: "Not a tree, bush or shrub of any kind but what bore unmistakable marks of the terrible fire which for weeks past had been going on." The last day of the month he jotted in his diary: "Large details, together with 650 Negroes, have been set to work on the inner lines, completing the redoubts and strengthening the position by connecting these with a heavy stockade of green timber. The enemy still extending to the left." That same day to the south, at Jonesboro, the engineer battalion filed

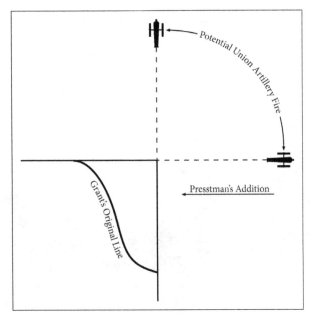

Fig. 8.4. The danger of angles. Presstman's addition to the Atlanta defenses roughly formed a right angle. This meant that Union artillery could potentially unleash deadly enfilade fire, shooting down the length of the trench. To protect from this scenario, defenders built traverses, which worked with varying degrees of success.

into the trenches behind Govan's Arkansas brigade. The subsequent battle proved a one-sided slaughter—2,200 Confederate casualties versus a mere 172 Federal. There is no record of engineer losses if any.[30]

August likewise proved a grueling month for Lieutenant Fergusson. He was ordered to report to Captain Coleman in Lee's Corps. On the 1st he noted: "Went over entire line and put in charge of all the sections of heavy work and others in charge of throwing the timber in front constructing abatis and chevaux-de-frise in the rear of the line. The main ditch was also worked upon." The lieutenant received orders from Captain Coleman on the 10th to sketch a map of all of the brigades and batteries in Lee's Corps by name, showing the topography, and "as well as you can" the position of the skirmish line and the position of the enemy's lines—especially the batteries. The next day he got a "fairly good survey" outside the lines, although he came under fire. He spent most of the day mapping indoors on the 17th, but later that day he was back on the line.[31]

On September 1, Frobel learned that the city was to be abandoned. That night he witnessed "the solid masses of infantry, marching in dead silence through the dark streets." Later that evening, he galloped ahead, hoping to

Fig. 8.5. Confederate Fort D. This 1864 photograph was taken from a casemate inside Atlanta Fort D following the capture of the city. Note the large beams to shore up the casemate and the sandbags around the embrasure. Courtesy Library of Congress, ppmsca32746//hdl.locpnp/p.print.

catch up with the rear guard. As he approached a hill, he heard an enormous boom—the ordnance train had been destroyed. Soon the explosions became continuous as car after car in domino effect exploded. He stood momentarily and watched the display, noting, "We had seen the last of Atlanta, and, sadly away, we once more put spur to our horses and galloped off."[32]

The Army of Tennessee's engineer corps reached its pinnacle in the Atlanta Campaign. While engineer officers prepared maps and staked off earthen positions that stretched for miles, evoking even the amazement and admiration of the enemy, the engineer troops and pioneers labored to both repair and detonate bridges, cut roads where none existed, corduroy the highways with saplings, and, as Hiram Williams told his diary, "make a lot of cheavaux-de-friz's." It could perhaps be argued that the horrid rains did as much to slow Sherman's advance as destroyed bridges and miles of trenches. Perhaps so, but the engineers had clearly done their part, reflecting an operational and tactical influence that surpassed their paltry numbers. As for Presstman, he had served both Johnston and Hood well, although neither would formally mention him in their after action reports.[33]

9

THE PONTONIERS

FOLLOWING THE LOSS of Atlanta, Hood fell back twenty-five miles southwest to Palmetto, Georgia. Sherman did not follow, and both armies settled in for a brief respite. Facing public and army consternation over the loss of the city, Hood placed the blame squarely on Hardee, due to his August 31 failure at Jonesboro. When Davis visited the army, Hood presented what Wiley Sword termed a "deceptively simple" offensive plan. He would avoid direct contact with Sherman's army and strike his extended supply lines in North Georgia, thus forcing the Federals to withdraw. Davis approved the plan, and then proceeded to make two major command changes. Hardee would be sent to the Department of South Carolina, Georgia, and Florida, thus making room for Cheatham at corps command. Additionally, the Army of Tennessee and the Department of Alabama, Mississippi, and East Louisiana, the latter now under Lieutenant General Richard Taylor, were combined to form the Military Division of the West, with Beauregard commanding. The position, as matters developed and as Beauregard would soon learn, proved nominal and Hood would largely ignore him.[1]

Changes also occurred in the engineers. M. L. Smith, who had been overseeing the defenses at Macon and Columbus, would be reassigned to Beauregard at department headquarters, with Frobel replacing him at Macon. Captain Winston, "one of the most skillful engineers in the service," was also employed at Macon. The works consisted of a series of lunettes, each capable of holding 200–300 men, and connected by a covered way and rifle pits. The unusual layout was due "to the genius of Gen. Smith, who designed them, and to the skill and energy of Capt. Winston." With the Army of Tennessee's top two engineers gone, command once again reverted to Lieutenant Colonel Presstman as "acting chief." An effort was made to fill the ranks of the 3rd Engineers with conscripts, but by August 31 the present-for-duty strength had dwindled to 388. As for Captain Lemuel Grant, he was transferred to oversee the Augusta defenses. With Macon possibly threatened, Hood also sent an ad hoc force, which included "Major Rowland's Engineer Company" and "Major Green's Engineer Company." All engineer troops were eventually sent back to the army.[2]

Captain Green was dispatched to build a wooden bridge across the Chattahoochee River at LaGrange, Georgia, to cross the wagon train, but he soon replaced it with a pontoon bridge. Throughout early and mid-October, Hood struck Sherman's communication, capturing Big Shanty and Acworth, but French's division was bloodily repulsed at Allatoona. The garrisons at Tilton, Dalton, and Mill Creek Gap were nonetheless easily bagged, and miles of track destroyed. "We have completely destroyed the enemy's railroad from Resaca to Tunnel Hill," Hood gloated on October 16. Sherman eventually called off the chase. He sent two of his corps to reinforce Tennessee, while another came from Missouri. With the balance of his army in Atlanta, he began his "March to the Sea."[3]

Department of Mississippi, Alabama, and East Louisiana

With Leadbetter out of the picture, probably for the best, engineer operations in the Department of Mississippi, Alabama, and East Louisiana reverted to Colonel Lockett. Most of the engineers in the department, like Lockett himself, were former Vicksburg prisoners, now paroled and exchanged. Lockett maintained his headquarters in Mobile, while he superintended three districts. The District of North and Central Alabama fell under Captain Powhatton Robinson, whose primary responsibility was the construction of earthworks at Montgomery. As of November 20, 1864, the works were only half finished, primarily because the slaves kept running away. The works at Selma were complete, Demopolis partially complete, and the Coosa River bridge complete. Captain Wintter oversaw the District of Mississippi and East Louisiana. His primary responsibility was keeping the pontoon bridges over the Pearl River at Jackson and the Tombigbee River at Columbus in service, and repairing and building of pontoon boats at Meridian. It was a huge district, and Wintter had only seven remaining members of his original sapper company, plus a small force of detailed men. In northern Mississippi, Lieutenant Donnellan superintended operations on the Yazoo River and maintained bridges at Grenada,

Table 9.1. Engineer Staff, Army of Tennessee, October 1864

OFFICER	ASSIGNMENT
Lt. Gen. Stephen W. Presstman	Acting chief engineer, Army of Tennessee
Maj. John W. Green	Staff, 3rd Engineers
Capt. Andrew H. Buchanan	Topographical engineer, Army of Tennessee
Maj. Arthur W. Gloster	Chief engineer, Cheatham's Corps
Maj. Thaddeus Coleman	Chief engineer, Lee's Corps
Capt. George M. Helm	Assistant engineer, Lee's Corps
Capt. John Vinet	Assistant engineer, Lee's Corps
Lt. W. W. Fergusson	Topographical engineer, Clayton's Division
Maj. George Pickett	Chief engineer, Stewart's Corps
Maj. Charles Foster	Assistant engineer, Stewart's Corps
Capt. Joshua A. Porter	Assistant engineer, Stewart's Corps
Capt. J. K. P. McFall	Assistant engineer, Stewart's Corps

Panola, and Abbeville. The District of the Gulf, which included the defense of Mobile, came under Von Sheliha. An engineer shop and depot was maintained at Demopolis, where wheelbarrows, tools, desks, tables, and drawing boards were made and repaired.[4]

Spanning the River

The Army of Tennessee arrived in Gadsden, Alabama, on October 20. In addition to Hood's 580 wagons, a soldier wrote of seeing 7,000 head of cattle and "a Pontoon train of 80 boats which are fitted as wagons like the body of a wagon. ... So you can see we are prepared for traveling." Beauregard arrived the next day and heard of Hood's new plan to move into Tennessee. The theater commander believed that the first target should be the Union supply depots at Bridgeport-Stevenson. He nonetheless saw certain advantages of moving into Tennessee, but he expressed concerns about logistics, of which "General Hood was disposed to be oblivious." He did insist that Wheeler's cavalry remain in Georgia to harass Sherman. Hood could retain "Red" Jackson's cavalry division, and Forrest would link up.[5]

Hood only remained for two days before moving west toward Guntersville. In his haste, and to Beauregard's utter disgust, he left behind the very thing that he needed the most—his pontoon train. The Creole had the bateaux removed from the Coosa River and hurriedly sent forward. Instead of crossing

Table 9.2. Department of Mississippi, Alabama, and East Louisiana

OFFICER	ASSIGNMENT
Col. Samuel H. Lockett	Department chief engineer
Capt. Powhatton Robinson	District of North and Central Alabama
Maj. Deitrich Wintter	District of Mississippi and East Louisiana
Lt. George Dunellan	District of North Mississippi
Lt. Col. Victor Von Scheliha	District of the Gulf

the Tennessee River at Guntersville, Alabama, as Hood had told Beauregard he would do, the army commander led the Army of Tennessee fifty miles west to Decatur, but it was guarded by 3,000 infantry and supported by two gunboats. Engineers suggested that a crossing could be made at Courtland, Alabama, but once there they considered it too risky. Hood thus continued west, this time planning to cross at Bainbridge, six miles above Florence. Bainbridge proved unfeasible, so Hood continued to Tuscumbia (a couple of miles south of the river from Florence), where the shoals protected a crossing both above and below. The shoals actually protected the Bainbridge crossing as well, but Tuscumbia was closer to the nearest railhead at Cherokee Station, Alabama. Using pontoons and makeshift paddles to launch an amphibious attack at Florence, the 1,000 Federal cavalry were easily chased off. Beauregard meanwhile was livid for Hood's failure to take him into his confidence.[6]

A bridgehead having been secured, the matter was now in the hands of the engineers. It would prove one of the largest challenges of the western engineer corps, and one for which it had prepared and trained since the previous April. The Confederate pontoon was built upon the model of the US Army, referred to as the "McClellan pontoon" or "Cincinnati pontoon," which in turn was copied from the French model. These were flat-bottomed boats 31 feet in length, 5 feet, 4 inches wide at the top, quite heavy, and requiring extra long flatbed wagons to haul them. The pontoons were placed 13 feet, 10 inches apart and lashed together by balks, long wooden poles 25.5 feet in length. Cordage was used throughout and no nails were applied. For a moderately sized river (about 445 feet), the typical French pontoon train required thirty-four bateaux, each drawn by its own wagon, thirty-two additional wagons (twenty-two for chess, eight for balks, chains, and cables, and two for tools, spare cordage, pickets to secure the cables, and pumps), and two traveling forges. Theoretically the Confederates would need over twice this train to cross the Tennessee River.[7]

The Confederates estimated the river at Florence to be 1,000 yards wide, but they may have underestimated. A modern map search of the Old Historic Railroad Bridge, built in 1840, reveals a river width of at least 1,500 feet. Initial estimates were that 135 boats (Leadbetter estimated 150) would be required. The Tennessee River at Chattanooga, which was 600 yards wide at that point, was in the process of having a pontoon bridge laid, and in that instance forty-seven boats were used, or about one boat every thirteen yards, the standard French pattern. By that measurement, the Florence crossing would have required about 115 bateaux. If the army actually possessed about eighty pontoons, as the Gadsden eyewitness claimed, then that number would have been far short. Back in February, Johnston claimed that 130 wagons were being altered to carry pontoons. Whatever the number, it appears to have been insufficient. The reserve pontoons at Meridian were delayed by Mobile & Ohio officials due to their sheer weight. The engineers resolved the shortage by rounding up "all manner of boats, rafts, and barges," according to an artilleryman. Close to a dozen flats were at Garner's Ferry, about four miles west of Florence; whether or not any of these were used to complete the project is not known. The stone piers of the destroyed railroad bridge still remained and helped to stabilize the bridge, but even so the strong current of the Tennessee River made for a "very imperfect and unsafe affair," bulging the bridge in the center to a crescent shape.[8]

The initial crossing of Lee's Corps on the night of October 30 through November 1 went smoothly. The brigades crossed in column of fours with bands playing. One soldier observed that it was the "longest Pontoon bridge I have ever seen," and at a distance it appeared as though "the men were walking on water." A sergeant in the Florida brigade thought that "the passage on the pontoons was a grand sight. Infantry columns—Artillery, immense wagon trains, Cavalry, droves of beef cattle, etc. etc. all to be seen from a lofty rock over looking the river presented a picture truly interesting and imposing." Unfortunately, all did not go as planned. There was a Yankee sabotage attempt made by a party floating upriver against the current and cutting the main guide rope with axes. The chain held, however, and three of the saboteurs were captured.[9]

M. A. Traynham, a member of Clarkson's Company F, 3rd Engineers, took a moment to write his wife once he arrived in Florence. He related how his company ferried their wagons and mules across the Tennessee River in pontoons and camped on a hill with a portion of the engineer regiment. On November 3, his company was sent into the countryside to collect timber for the pontoon decking, an indication that the pontoon train either lacked or had insufficient chess. On the 9th, the men were "guarding the bridge and building boats."

Two days later a detail from the company went with Captain Clarkson to Tuscumbia to secure lumber. Traynham became sick as the weather turned increasingly cold.[10]

Cheatham's infantry marched across on November 9. As the infantry cleared, a herd of crossing cattle caused a stampede, breaking the bridge and sending eighty head and several men into the Tennessee River. The pontoons were repaired, but a subsequent rainstorm broke the bridge once again; all crossings were suspended for four days. Passage resumed on the 13th, with Lieutenant Green of the engineers telling his mother: "The timely arrival yesterday of a supply of pontoon boats from Demopolis added to ours" enabled the Confederates to "span a river nearly ½ mile wide." Nonetheless, according to a Georgian, the rapid current continued to place a great strain on the bridge. By the 16th the *Chicago Tribune* reported: "The Rebels have a good bridge [at Florence] from one shore to the other." Beauregard desired to know the dimensions of the planks and timbers needed for a more permanent bridge once the army departed. He inquired if the boats should be built in Selma or Corinth. The officer in Atlanta who oversaw the construction of pontoons (probably Arthur Gloster) could perhaps be spared to oversee the current project. General Taylor was ordered to immediately send all available bateaux to Corinth.[11]

There was also the issue of keeping the railroad running. The supply chain now emanated from Montgomery, Selma, Mobile, and Demopolis, with all rations, forage, ammunition, medical supplies, and pontoons being shipped north along the Mobile & Ohio, then operating twenty-seven carloads a day, to Corinth and then east along the Memphis & Charleston Railroad to Florence. As always there were problems. The Memphis & Charleston was out of commission for fourteen miles, between Cherokee Station and Tuscumbia, due to destroyed bridges and washed-out culverts. How long it would take to repair the breach was anyone's guess. If all of the needed tools, including shovels, cross saws, picks, augers, and axes could be furnished to cut and saw the trestle beams and crossties, then three-fourths of a mile per day would be optimal, meaning three weeks. All impressed Blacks were then at work on the Mobile & Ohio from Okolona north, keeping it functioning. Additional labor would have to be impressed Blacks, a difficult task in North Mississippi by that time. Two of Presstman's engineer companies were thus requested, one for each railroad. Only eleven of the fourteen miles of the Memphis & Charleston were ever completed. Smith was also charged with building blockhouses and other small earthen works for one or more companies along both railroads.[12]

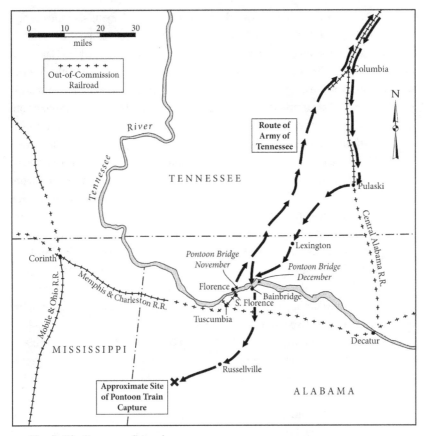

Map 5. The Tennessee Campaign.

In addition to laying pontoon bridges and keeping the railroads operational, the engineers also had responsibility for preparing fortifications, the foremost of which was at Corinth, the army's new supply depot. Major General Frank Gardner, a West Pointer with long Old Army experience, was in command at that location, and he was not pleased at the progress. He had only one engineer officer, Lieutenant S. McD. Vernon, at his disposal. He had previously ordered Major Meriwether to Corinth, but while en route at Okolona, Mississippi, the officer became ill. M. L. Smith and Lockett were both present, but a dispute soon developed. Gardner's orders were to construct defensive stockades at Corinth and along the railroad, but the engineers desired to use and even expand upon some of the old Corinth fortifications, which Beauregard had declared as too extensive even for a 40,000-man army. "With all due respect to their higher position and higher talent," Gardner disgustedly wrote, "I think it is absurd to attempt to defend this depot over a space of about six miles or upwards, but it is not in my power to order either of them to change their plans."[13]

Wintter recuperated and belatedly reported for duty. He resolved Gardner's concerns about the overly expansive Corinth fortifications by enclosing the salient points of the old Union earthworks and repairing the detached works, thus making an accommodation for a more reasonable 1,000-man garrison; even scraping together that many men would be problematic. Lockett meanwhile continued his reconnaissance along the Tennessee River, from Pride's Ferry, Alabama, upriver almost to the old Shiloh battlefield. A battery had to be constructed somewhere along this line to prevent enemy gunboats from venturing too close to Florence and Hood's lifeline bridge. Chickasaw Bluffs, Alabama, ten miles above Eastport, Mississippi, was selected for a four-gun Parrott rifle battery. An old Indian mound had been converted into a battery position by the Confederates back in the spring of 1862. It was a feeble position, hardly capable of stopping an ironclad, but it was something. Engineer troops kept the crossings at Cheatham's and Garner's Ferries in operating order. Work meanwhile continued on the Memphis & Charleston Railroad breach, the work being done by two companies of engineer troops and a force of Blacks.[14]

With work underway on the Corinth fortifications and the Memphis & Charleston repairs, the issue of pontoon construction now had to be addressed. Twenty boats were on their way from Columbus, Mississippi, to Panola, and these were diverted to Presstman to complete his train that would accompany the army into Tennessee. Beauregard ordered a reserve pontoon train built, fifty of the boats to be made at Demopolis and fifty at Mobile. The necessary seasoned timber could be easily cut at Mobile, and by November 15 some twenty bateaux had been prepared, with two boats being made a day. The Demopolis office struggled with seasoned timber, all of the lumber having to be first kiln-dried. Construction did not begin until December 1, but the frameworks went quickly and the fifty boats were reported to be on schedule. Arrangements were made with the quartermaster depot at Meridian for the necessary balks and chess. The timber was prepared, but for some unknown reason the lumber was not shipped before the Mobile & Ohio was shut down.[15]

Hood belatedly advanced his army from Florence on November 21, with Forrest's cavalry in the van. "I never was so hopeful for our immediate success than now," wrote Colonel Walker. Harsh realities soon set in—inch-thick ice, nearly impassable muddy roads, poorly clad troops, and slim rations. John M. Schofield's Federal army of 23,000 infantry and James Wilson's 7,000 cavalry were known to be at Pulaski, but as the Confederates approached, the bluecoats withdrew thirty miles north to Columbia on the Duck River. On the night of the 27th, Hood established his headquarters three miles south of Columbia. The pontoon train, boats being drawn by oxen, trailed a day's march behind. The general planned to get in the enemy's rear, but for that he needed a pontoon bridge. Fortunately, a young scout in the 1st Tennessee, G. Wash Gordon, told Presstman about Davis Ford, four miles east of town, which was unguarded by the Federals. The engineers went to work grading an approach while awaiting the bateaux. The lumbering pontoon train moved east on East Ninth Street, and could easily be observed as it passed the enemy-held ridge north of town. Throughout the night of the 28th, the engineers labored to get the bridge into position.[16]

The next morning, while S. D. Lee's Corps shelled the town and occupied the Federal's attention, the balance of Hood's army crossed and headed for Spring Hill, ten miles north, but failed to block Schofield's retreat. The subsequent Battles of Franklin (November 30) and Nashville (December 15-16) left the Army of Tennessee if not destroyed at least damaged beyond repair, losing 13,500 troops, and leaving Hood with only about 18,800, with Thomas's 40,000 in pursuit, not to mention a 5,000-man Federal division closing in from Decatur.

Re-Crossing the Tennessee

Following Hood's rout at Nashville, Thomas launched what Hess termed "one of the largest pursuits of a beaten army during the Civil War." From Nashville to the Alabama line was ninety miles, crossed by the Harpeth and Duck Rivers and numerous creeks, all of which had bridges destroyed by the Confederates. The speed of the Federal vanguard—Major General James H. Wilson's cavalry corps of nine brigades and Brigadier General Thomas Wood's IV Corps, was a testament to northern ingenuity. The bluecoats improvised rafts and left their artillery and wagons behind, all in an effort to cut off Hood's retreating army before it could cross the Tennessee River. The Union pontoon train, stymied by heavy rains, finally caught up with the IV Corps on the morning of December 22 at the Duck River. The army crossed for the next two days, with the flimsy bridge breaking several times under the pressure of the rushing water.

Between fifteen to twenty of Hood's ordnance wagons were abandoned so that his pontoons could be double-teamed. Even so, there were at least fifteen broken-down bateaux stuck in the nearly impenetrable snow and mud. A detail of 200 men was sent back to assist the engineers in retrieving the boats. The army was still eleven miles from the Tennessee River on December 24, but Wilson's cavalry was thirty-seven miles to the northeast. "A Mr. Carter says the colonel commanding the pontoon transportation told him he was going to

Bainbridge," Wilson reported. Thomas still held out hope. It all depended on how fast Hood could build his pontoon bridge and cross his ragged army. In a rare bit of luck, Hood had previously dispatched Robert Cobb's pontoniers to Decatur to retrieve fifteen captured pontoons that had fallen into the hands of Phillip Roddey's cavalry brigade. Precisely how the boats were left when the Federals abandoned the town was never determined. Apparently there was a miscommunication that the Union gunboats would tow them to safety, but due to "some blunder" they were left. Major Pharr believed that the escape of the Army of Tennessee from Thomas's pursuing divisions now came down to Cobb's men and how quickly they could get the pontoons downstream to Bainbridge, Alabama, where the army would be crossing.[17]

On Christmas Eve, Hood sent a dispatch to Lieutenant General Stewart stating that before daylight all of his pioneer parties should be sent to Presstman at Bainbridge. "Have all your empty wagons filled with plank for decking, gathering it from buildings on the road, let them go to Presstman also." At 6:20 p.m. that same day, Stewart again received an urgent dispatch, noting that "if Major Green, of the engineers, catches up to you give him the road and let him pass with his train of decking, etc. He is expected to be in Lexington [Alabama] this evening with the rear of his train." Christmas Day dawned gray, drizzly, and bitterly cold as the engineers began laying the pontoon bridge across the shoals "as rapidly as the arrival of the boats would allow." The crossing was not a regular ferry, and the river was quite rapid. Indeed, the site had been rejected as a crossing back in November. At the most opportune time, Cobb's pontoniers were seen floating down the river with the captured Decatur pontoons. A wild cheer arose from the army—"Cobb has come! Cobb has come!" The bridge was completed by dawn of the 26th. The strong current bowed the bridge at the center, and only a few guns or wagons could cross at a time. Cheatham's and Lee's Corps crossed on the 26th and 27th, and Stewart's Corps and the cavalry the next day. The boats were taken up on the 28th.[18]

The Federals hoped that the navy could accomplish what had eluded the army. Admiral Samuel P. Lee had two light-draft gunboats west of Florence. If they could punch their way past a rebel battery on the north bank, at the foot of the shoals, about six miles from Hood's pontoon bridge, then perhaps the remnants of the Confederate army could yet be trapped on the north bank. Lee believed that the southerners could not cross at Bainbridge, Little Muscle Shoals as it was called, unless the river dropped. "Hood must have been sorely pushed to have resorted to [cross at] such a place as the shoals," he wrote Thomas. The river was falling, but what was good news for Hood was bad news for Lee; his boats could ascend no farther. Nonetheless, the brief booming of

Fig. 9.1. Federal pontoon bridge at Decatur, Alabama, in 1864. Courtesy Library of Congress, LC-USZ62-104302.

gunfire between the boats and a four-gun rebel battery could be heard in the distance, adding to a nerve-racking situation at Bainbridge.[19]

Once safely over the Tennessee River, the main army proceeded to Tuscumbia and on to Corinth. At least eighteen of the pontoons went with the army. A detail under A. J. Godwin was ordered to place a pontoon bridge across Big Bear Creek. Once the troops and wagons had crossed, the boats were destroyed. The main pontoon train, along with a number of supply wagons, proceeded to Russellville, Alabama, on its way to Columbus, Mississippi. At dusk, about seven to ten miles (the accounts vary) out of Russellville on the road to Nauvoo, the 15th Pennsylvania Cavalry, in the van of a column of Union cavalry, pounced on the train. The panic-stricken teamsters cut the best mules from their harnesses and struck out for Columbus. Union reports tallied 78 pontoons and 200 wagons captured; the Confederates claimed 83 pontoons and 150 wagons captured. All accounts agree that the train stretched for some five miles along the narrow road. One of the Pennsylvania troopers noted that the pontoons were "the finest I had ever seen, and most of them had the names of prominent southern ladies painted upon them, such as Lady Davis, Lady Bragg, etc." A Black man approached a Union officer from the woods and made a request: "Cap'n Gloster says he would like to have his papers." The officer asked where Gloster was. "Oh, he's jus out dar in the bushes." He was told that if Captain Gloster wanted his papers then he would have to come in and personally retrieve them; Gloster never made an appearance.[20]

The shattered remains of the Army of Tennessee would eventually transfer to North Carolina, where it joined an ad hoc army assembled under Joseph E. Johnston, but it proved too little, too late. The story of the engineer corps of the Heartland would thus seemingly come to an ignominious closure, but for one more tragic event. In early 1865, Presstman passed through Macon on his way to Richmond, collecting the most important maps to be stored at the Confederate War Department. While en route to Richmond through North Carolina, he ran across the track of the Piedmont Railroad, stumbled, and was killed by a locomotive. He thus joined the list of other fallen Heartland engineers—Gray, Dixon, Harris, and Wampler.[21]

Epilogue

THE WAR WAS OVER and the engineers returned to their prewar civilian pursuits. The wrecked South needed rebuilding, and engineers would again be in demand. Most of the Heartland engineers would live to see their profession undergo incredible transformation with the use of steel—a bridge spanning the Mississippi River, the world's first skyscraper (the Home Insurance Building) in Chicago, the advent of steel battleships, and a massively increased rail system. The profession also saw the beginning of engineering societies, including the prestigious American Society of Civil Engineers, of which several of the former Confederates, including Robert Cobb and William D. Pickett, became members. The Heartland and the nation were rapidly changing, and the engineers would continue to play a pivotal role.

One of the longest living of the engineer officers was George M. Helm. After the war he returned to Washington County, Mississippi, where he had previously been employed. He served as sheriff during Reconstruction, and later became one of the largest land owners in the Delta. He died in 1930 at the age of ninety-two. Leon Fremaux returned to New Orleans but never enjoyed large financial success. He became an honorary lieutenant colonel in the Louisiana National Guard until his death in 1898.[1]

Perhaps the most adventuresome lives were those of William D. Pickett and John Haydon. Returning to work with the Memphis & Ohio Railroad following the war, Pickett ventured out West in 1876, settling in Wyoming in 1883. He would serve three terms in the Wyoming state legislature before returning to Kentucky, where he died in 1917. It was said of him that he had "experience hunting grizzly bear greater probably than any man who ever lived." John

Haydon worked as a civil engineer for railroads in Maryland, Minnesota, Montana, and North Dakota. On several occasions while performing surveys he had to dodge hostile Indians. He died in 1902.[2]

Brigadier General Jeremy F. Gilmer, who served under Albert Sidney Johnston, was wounded at Shiloh, and ultimately headed the Engineer Bureau in Richmond, returned to Savannah after the war, where he so longed to be during the conflict. In 1867 he became president of the Savannah Gas Light Company, a position that he held until his death in 1883. Tragically, while attending a Drainage Committee meeting one evening, he was thrown by his horse, which left him disabled for the balance of his life.[3]

Victor Von Sheliha, who served in both West and East Tennessee before transferring to Mobile, would resign his commission in August 1864 while the city was under siege, prompting Gilmer's complete contempt. The German-born engineer went to New Orleans after the war, dying in 1903 at the age of seventy-seven. Lemuel P. Grant became president of the Western & Atlantic Railroad. His house still stands in Atlanta, and the 100 acres that he donated to the city (Grant Park) today houses the Cyclorama and the Atlanta City Zoo. Samuel H. Lockett taught at East Tennessee University and Louisiana State University at the conclusion of the war, before becoming a colonel in the Egyptian Army, and later serving as a railroad contractor in Chili. He died in Bogota, Columbia, in 1891.[4]

Bushrod W. Frobel, who briefly served as the chief engineer for the Army of Tennessee during the Atlanta Campaign, became superintendent of public buildings for the state of Georgia, supervising work on the capital, governor's mansion, penitentiary, and mental institute. He eventually was appointed US assistant engineer in Georgia, directing work on the state's rivers and harbors. He ended his career as general manager and chief engineer of the Macon & Covington Railroad before his death in 1888 after a brief illness.[5]

Most of the engineers lived routine lives, doing the same work they had done prior to the war—Henry N. Pharr in Arkansas, Andrew H. Buchanan teaching engineering in college, and Arthur Gloster in East Tennessee. Robert Cobb became a civil engineer for several railroads, which carried him to different states. He eventually became chief engineer of the Ohio & Southern Railroad in Cleveland, Ohio. In 1895, at the age of fifty-five, he succumbed to a flu-like disease. Wilbur Foster, the young officer who so defiantly opposed the location of Fort Henry, became an engineer for the Nashville Water Works. When Foster retired in 1912, he purchased a two-row typewriter and pecked out his autobiography, based on the diary he kept during the war. The manuscript subsequently went to his daughter, but was destroyed when her house later went up in flames. He lived until 1922, never taking a vacation. When told by the doctor that he would have to reduce his cigar smoking to one a day, he had eight-inch-long cigars specially made. He died a charter member of the Engineering Association of the South.[6]

The Confederate engineers of the Heartland were constantly stressed by the pressures of time. Short on personnel, they were forced to travel huge distances stomping out brushfires. Under the circumstances, it is not surprising that their work did not prove innovative or transformative. Nonetheless, their contribution should not be dismissed. But for the engineers, Hood's 1864 Tennessee Campaign could not have been conducted, nor would the railroad bridges in East Tennessee have remained operational. The Island No. 10 defenses, flawed though they were, proved sufficient to hold at bay John Pope's 20,000-man Federal army for five weeks, thus buying time for Albert Sidney Johnston to potentially score a victory at Shiloh. In the Vicksburg Campaign, it was leadership and strategy that failed, not Lockett's engineering. The war was now concluded and daunting challenges lay ahead. It was time for the engineers to go back to work.

APPENDIX A

Engineer Officers of the Heartland

HOW MANY ENGINEERS served in the Heartland? The answer, of course, depends on how the term "engineer" is defined. Does the number include only professionally trained military engineers who served in the US Corps of Engineers? Could the number include civil engineers with no formal military training? Does the number embrace all officers, including those who, as Solonick stated, "simply oversaw military tasks"? The categories below attempt to answer these questions.[1]

1. *West Pointers involved in engineering projects who had either previously served in the US Corps of Engineers or US Topographical Engineers, or who had taught engineering at the academy:* E. P. Alexander, P. G. T. Beauregard, Lewis DeRussy, Joseph Dixon, David B. Harris, Jeremy Gilmer, Danville Leadbetter, Samuel H. Lockett, Martin L. Smith.

2. *West Point graduates who served in the US infantry, artillery, or cavalry and who perfunctorily advised on Confederate engineering projects:* James Bushrod Johnson (taught civil engineering), Leonidas Polk, Francis Shoup, Philip Stockton, Lloyd Tilghman (civil engineer), John Villepigue.

3. *Non-West Point graduates who perfunctorily advised on engineering projects:* Adolphus Heiman (foreign-trained engineer), Gideon Pillow.

4. *West Point graduates who became Confederate engineer officers:* Achilles Bowen, James M. Couper, Henry De Veuve, Bushrod W. Frobel, John Pegram.

5. *Prewar civil engineers who served as officers or under contract as civilians:* Adna Anderson, Andrew H. Buchanan, John J. Clarke, A. W. Clarkson, Robert L. Cobb, Thaddeus C. Coleman, Frederick Y. Dabney, Amos S. Darrow, James J. Davies, Arthur B. De Saulles, Amory Dexter, George Donnellan, Thomas L. Estill, Henry C. Force, Wilbur F. Foster, Leon J. Fremaux, William Freret, T. J. Glenn, Arthur W. Gloster, Lemuel P. Grant, George H.

APPENDIX A: ENGINEER OFFICERS OF THE HEARTLAND

Hazlehurst, John W. Green, John Haydon, George M. Helm, James H. Humphreys, John G. Kelly, Montgomery Lynch, John G. Mann, George R. Margraves, Richard McCalla, Minor Meriwether, Thomas Jefferson Moncure, John Moore, Walter J. Morris, T. S. Newcomb, Henry N. Pharr, George B. Pickett, William D. Pickett, Joshua A. Porter, Stephen W. Presstman, William D. Prinz, W. A. Ramsey, John M. Robinson, Powhatton Robinson, Robert P. Rowley, Edmund W. Rucker, Edward B. Sayers, Felix R. R. Smith, John F. Steele, Silvanus W. Steele, James D. Thomas, John S. Tyner, Rush Van Leer, G. O. Watts, Waightsill A. Ramsey.

6. *Architects who served as engineer officers:* Calvin Fay, Deitrich Wintter.

7. *Foreign-trained engineers who served as Confederate engineer officers:* James Norquet, Victor Von Sheliha, Francois J. Thysseus.

8. *US Coast Topographical Survey officers who became Confederate engineer officers:* John M. Wampler, Henry Ginder.

9. *Non-civil engineers previously in the employ of a railroad who became engineer officers:* James K. P. McFall.

10. *Non-professionals of known occupations who served as engineer officers:* R. H. Armstrong (farmer), William W. Fergusson (teacher), Jules V. Gillimard (chemist), William A. Hansell (military student), Menefee Huston (college student), D. H. Huyett (illustrator), E. McMahon (machinist), Joseph A. Miller (foundry owner), Thomas Jefferson Ridly (machinist), John Baptist Vinet (New Orleans businessman).

11. *Surveyors who served as engineer officers:* Andrew B. Gray, James Hogane, Matthew Fontaine Maury Sr.

12. *Engineer officers with unknown prewar occupations:* James H. Allen, Arthur S. Barnes, P. J. Blessing, G. H. Browne, J. J. Conway, D. W. Currie, Francis Gillooly, C. C. Gooden, A. W. Johnson, W. A. C. Jones, G. W. Maxson, George R. McRee, S. W. McVernon, James S. Morrison, Richard A. O'Hea, Henry Otey Minor, John Palmer, R. Riddlely, B. F. Roberts, J. Hudson Snowden, P. W. Summer, Napoleon B. Winchester.

13. Engineer officers who served either exclusively or near exclusively as draftsmen or with the topographical department (antebellum occupations given when known): E. G. Anstey (musician), Charles Foster (illustrator), Frank Gaines, Fred G. Gutherz, James H. Hainey (civil engineer), J. F. Knight, Lee Mallory, Albert Martin, Conrad Meister, Valentine Herman, J.

APPENDIX A: ENGINEER OFFICERS OF THE HEARTLAND

W. McQuire, H. A. Pattison (civil engineer), R. R. Southard, John S. Stewart, J. Lois Tucker.

14. *Civil engineers serving in 3rd Engineers as noncommissioned officers:* Charles R. Boyd.

15. *Graduates of a foreign military school:* T. G. Raven.

16. *Graduate of a southern military school:* Miles M. Farrow, Charles F. Baker (draftsman).

Of the 139 traceable engineers and draftsmen who served in the Heartland, 61 percent were either prewar professional engineers or had received some engineering training in military or foreign schools.

APPENDIX B

Occupations in the 3rd Engineers

THE 3RD ENGINEER REGIMENT, like all Civil War regiments, had hundreds of men filter in and out of the ranks. There were two classifications—artificer and laborer, but neither of these offered specific prewar occupations. Unfortunately, only sixty-two privates and noncommissioned officers were found to have specific occupations listed. Nonetheless, this pool offers a sampling of those who served. It must be remembered that under the Confederate law, half of the engineer troops could be skilled and half nonskilled. The average age of both groups was thirty.

Nonskilled Occupations	
Farmer	21
Merchant	10
Clerk	2
Tailor	1
Student	1
Railroad laborer	1
Laborer	1

Skilled Occupations	
Mechanic	11
Engineer	1
Stone mason	1
Carpenter	11
Blacksmith	1
Wheelwright	1

Glossary

abatis: A defensive barricade of cut trees with the sharpened branches facing the enemy.
balks: Longitudinal timbers connecting pontoon boats.
bastion: A structure projecting from a fortification.
bombproof: A shelter of wood and earth to protect against artillery fire.
chess: Floor planks of a pontoon bridge.
chevaux-de-frise: A horizonal beam with pointed stakes set at different angles.
cremalliere line: A zig-zag-shaped defensive earthen fortification.
embrasure: An opening in a parapet.
en barbette: A gun firing over a parapet.
hachures: Short, even-spaced parallel lines used on a topographical map to indicate slopes.
lunette: A defensive fortification shaped like a half-moor or crescent.
military crest: Position on the slope of a hill offering the most direct line of fire.
redan: A V-shaped defensive fortification with the apex pointed toward the enemy.
redoubt: An enclosed defensive fortification built either inside or outside a larger fortification. It is typically square- or pentagonal-shaped.
re-entrant angle: An angle pointed away from the enemy.
sappers and miners: Technically, sapper and miner companies were men trained to dig trenches with sap rollers at the head of the trench. In the Confederacy, however, the term was loosely used, with sapper companies being the forerunner to engineer troops.
sap roller: A large, circular basket made of horizontal sticks (stakes) woven with vertical saplings (rods) and packed in the center with bundles of sticks and dirt. When turned on its side, the basket (gabion) could be rolled along a trench (sap) toward the enemy, offering protection from enemy fire.
stockade: A vertical wall of pointed stakes, similar to a western frontier fort.
tete-de-pont: A defensive work at the end of a bridge.
traverse: A mound of dirt inside a trench or fortification running perpendicular to it so as to protect from explosives or enfilade fire.

Notes

Abbreviations

ADAH	Alabama Department of Archives and History, Montgomery, Alabama
AHC	Atlanta History Center, Atlanta, Georgia
AU	Auburn University, Special Collections, Auburn, Alabama
CED	Letters and Telegrams Sent, Confederate Engineer Department, 1861–1864, RG 109, National Archives, Washington, D.C.
CMSR	Compiled Military Service Records, National Archives, Washington, D.C.
DU	Duke University, David M. Rubenstein Rare Book and Manuscript Library, Durham, North Carolina
EDLS	Engineer Department, Letters and Telegrams Sent, 1861–1864, M628, National Archives Record Group 109, Washington, D.C.
ETSU	East Tennessee State University, Johnson City, Tennessee
EU	Emory University, Manuscripts, Archives, and Rare Book Library, Atlanta, Georgia
GLC	Gilder Lehrman Collection, Chestertown, Maryland
LSU	Louisiana State University, Baton Rouge, Louisiana
MDAH	Mississippi Department Archives and History, Jackson, Mississippi
MPL	Memphis Public Library, Main Branch, Memphis, Tennessee
MVC	Mississippi Valley Collection, University of Memphis, Memphis, Tennessee
NPS	National Park Service, various parks
OR	*War of the Rebellion: A Compilation of the Official Records of Union and Confederate Armies,* 128 vols. Washington, D.C.: US Government Printing Office, 1880–1901. Unless otherwise cited, all references are in Series 1.
ORN	*Official Records of the Union and Confederate Navies in the War of Rebellion,* 30 vols. Washington, D.C.: US Government Printing Office, 1894–1927.
SHC	Southern Historical Collection, University of North Carolina, Chapel Hill, Chapel Hill, North Carolina
TSLA	Tennessee State Library and Archives, Nashville, Tennessee
TUT	ulane University, New Orleans, Louisiana
UK	University of Kentucky, Special Collections Research Center, Margaret J. King Library, Lexington, Kentucky
UVA	University of Virginia, Albert and Shirley Small Special Collections Library, Charlottesville, Virginia
WRHS	Western Reserve Historical Society, Cleveland, Ohio

Preface

1. Hess, "Revitalizing Traditional Military History," 20–33.
2. georgiaencyclopedia.org/?s=appalachian+province; store.usgs.gov/assests/MOD.StoreFiles/Ecregion/21632_tn.front_.pdf.
3. Hess, *Field Armies and Fortifications*, 17; Allardice, *Confederate Colonels*, 324; Rhodes, "Jeremy Gilmer," 94–113.

1. Defending the Mississippi River

1. Woodworth, *This Great Struggle,* 58; McMurry, *Two Great Rebel Armies,* 14–15; Hess, *Civil War in the West,* 2–3. Murray and Hseieh (*A Savage War,* 63) declared the Mississippi River as "the great prize in the West."
2. Trautmann, ed., *A Prussian Observes,* 153, 185.
3. Daniel and Bock, *Island No. 10,* 12–13; Gilmer to Seddon, December 26, 1862, *OR,* Series 4, 2:259–60.
4. Hughes and Stonesifer, *Pillow,* 8–104, 157, 162, 319–29, 324–25; Walker to Harris, April 19, 1861, Harris to Walker, May 8, 1861, and Pillow to Walker, May 9, 1861, *OR,* 52 (2): 56, 90; *Memphis Appeal,* April 27, 1861.
5. Pillow to Walker, April 20, 1861, Walker to Pillow, April 20, 1861; Cooper to Stockton, April 22, 1861; Harris to Walker, April 22, 1861; Tate to Walker, April 23, 1861; and Pillow to Walker, April 26, 1861, *OR,* 52 (2): 57–58, 62–63, 67, 72; Allaradice, *Confederate Colonels,* 358; Solonick, *Engineering Victory,* 3, 4, 13–14, 21; Abbott, *Memoir of Dennis Hart Mahan,* 34–36; Hess, *Field Armies and Fortifications,* 1–2; Hess, *Fighting for Atlanta,* 8–9.
6. "Biographical Sketch of Colonel William D. Pickett," UK; *Memphis Appeal,* April 19, 1861; American Society of Civil Engineers, *Transactions of the American Society of Civil Engineers* vol. 81, 1671; *Williams' City Directory 1860,* 34, 250, 275; Mitchell, *Tennessee State Gazetteer and Business Directory,* 149, 341; findagrave.com/memorial/18429/minor-meriwether; Calvin Fay, CMSR; Pickett, "The Bursting of 'Lady Polk,'" 277; Krick, *Staff Officers in Gray,* 344; Lindsley, ed., *Military Annals of Tennessee,* 879–80; Allardice, *Confederate Colonels,* 329; Edward McMahon, CMSR. Wintter was a partner in the architectural firm of Fletcher & Wintter. He settled in Memphis after the war and died in 1872.
7. www.weaverassociates.com/experience/fort_pickering.html, accessed March 3, 2020; Pillow to Stockton, May 28, 1861, Gideon Pillow Papers, MPL; *Memphis Appeal,* June 6, 1861; Russell, *My Diary North and South,* 307.
8. *Memphis Appeal,* April 29, May 5, June 6, 1861; *New York Herald,* December 4, 1861.
9. Lee to Walker, May 10, 1861, *OR,* 52 (2): 93; *Williams City Directory,* 17; *Memphis Appeal,* June 6, 1861; 1860 US Census (Lynch age). Lynch described himself thus: "I am a native of Petersburg, Virginia, received a military education [not V.M.I.] and have been in the army; engaged for 25 years constructing some of the most important works in the South, most of the time as chief engineer." Montgomery Lynch, CMSR.
10. Daniel and Bock, *Island No. 10,* 4; Pillow to Walker, May 9, 1861, *OR,* 52 (2): 90; Connelly, *Army of the Heartland,* 30–31, 40.
11. *New York Herald,* December 4, 1861; *Chicago Tribune,* June 10, 1862; Beauregard to Pickney, April 24, 1862, *ORN,* 23:698; Field, *American Civil War Fortifications,* 18; Russell, *My Diary*

North and South, 309–13; Hughes and Stonesifer, *Pillow,* 171. An earthen powder magazine (on private property) remains to this day.

12. Hughes and Stonesifer, *Pillow,* 162; Stewart to Jordon, March 21, 1861, *OR,* 10 (2): 352; Lynch to Polk, December 1, 1861, *OR,* 7:728; *New York Tribune,* June 12, 1862; Field, *American Civil War Fortifications,* 17. Fort Pillow is today a 1,642-acre state park, and much of the outer defensive line remains in a remarkable state of preservation.

13. Horn, *Leonidas Polk,* 168–71; Hughes and Stonesifer, *Pillow,* 176–77, 182–84.

14. Andrew Belcher Gray, Biographical Summary, ms 1250, Arizona Historical Society, Tucson. The Lake Superior maps are in TSLA, acc. no. 2002-070, ID 29579. In Civil War documents, Gray signed his name as "Asa Gray." The reason behind this remains a mystery.

15. Daniel and Bock, *Island No. 10,* 4; Polk, *Leonidas Polk,* 2:17; Trautman, ed., *A Prussian Observes,* 159; *Chicago Tribune,* April 12, 1862; *New York Herald,* April 9, 1862; *New Orleans Delta,* March 8, 1862. Island No. 10 is today part of the Missouri shore. Tiptonville, which during the war was a river town, is now a mile from the river.

16. Daniel and Bock, *Island No. 10,* 4; memorandum on the condition at Island No. 10, *ORN,* 22:747; Pillow to Polk, August 28, 1861, *OR,* 3:685; Redan Fort, Camp Polk, Island No. 10, May 27, 1862, *OR,* 8:142; findagrave.com/memorial/153606352/andrew-belcher-gray, accessed March 17, 2020.

17. Gray to Blake, September 18, 1861, *OR,* 3:703–5; J. Hudson Snowden, CMSR; Polk, *Leonidas Polk,* 2:44; Rowley to R. W. Johnson, September 25, 1862, and J. P. Champenas to Rives, September 10, 1862, Robert P. Rowley, CMSR. Although listed as a topographical engineer, there is no evidence that Snowden was a civil engineer. He was apparently a Texan detailed from the infantry as an engineer. See J. Hudson Snowden, CMSR. Polk later referred to Snowden as "one of my aids" [*sic*].

18. "List of Fortifications in the State of Tennessee," September 15, 1861, *OR,* 4:408.

19. Hess, *Civil War in the West,* 9; Woodworth, *Jefferson Davis and His Generals,* 37, 39–41; Horn, *Leonidas Polk,* 179–80.

20. Abstract from return, September 30, 1861, *OR,* 3:712; Albert Fielder Diary, September 8, 1861, TSLA; Johnston, *Life of Gen. Albert Sidney Johnston,* 324. See also James I. Hall to brother, September 25, 1861, James I. Hall Letters, SHC; Rucker to Polk, September 22, 1861, Edmund Rucker, CMSR.

21. Welker, *Keystone Rebel,* 66; Johnston, *Life of Gen. Albert Sidney Johnston,* 308, 318, 324, 325.

22. Johnston to Cooper, September 25, 1861, Cooper to Johnston, September 25, 1861, Cooper to Johnston, September 26, 1861, and Benjamin to Cooper, September 27, 1861, *OR,* 4:426–27, 429, 430.

23. Polk to Walker, August 3, 1861, and Benjamin to Polk, November 4, 1861, *OR,* 4:380, 508–9; Steven M. Mayeux, "The Life and Times of Lewis DeRussy," http://fortderussy.org/derussy-bio.html, accessed March 14, 2020; Allardice, *Confederate Colonels,* 127; Miller to Polk, December 1, 1861, A. P. Stewart to Polk, November 9, 1861, and Miller to Polk, December 28, 1861, J. A. Miller, CMSR; findagrave.com/memorial/94356925/joseph-a-miller.

24. Trezevant to Polk, November 21, 1861, *OR,* 52 (2): 215; Baker, "Island No. 10," 55; Frost to Cabell, December 9, 1861, *ORN,* 22:808; "William D. Pickett," in American Society of Civil Engineers, *Transactions of the American Society of Civil Engineers,* 81:1671.

25. Hughes, *Battle of Belmont,* 60–64; Polk, *Leonidas Polk,* 2:44–45; A. G. G., "The Bursting of 'Lady Polk,'" 118–19; Pickett, "The Bursting of 'Lady Polk,'" 278; *New Orleans Picayune,* November

14, 1861. About 10 percent of the original Confederate earthworks survive at the Belmont State Park in Columbus. Half of Fort DeRussy has eroded.

26. Meriwether to Polk, December 20, 1861, Minor Weriwether, CMSR.

27. Hughes, *Battle of Belmont,* 56–57; *Chicago Tribune,* March 6, 1862; *New York Herald,* March 6, 1862; Stevenson, *Thirteen Months in the Rebel Army,* 65–66; US War Department, *The Official Military Atlas of the Civil War,* plate V.2; fortderussy.org/otherforts.html, accessed March 15, 2020. In the 1920s the town, due to devastating floods, was moved atop the bluff. Polk claimed 140 guns at Columbus, but a document showed sixty-one heavy guns, twenty-three being either large caliber or rifled. The precise number of field guns is unknown. "Cannon Inventory at Fort Columbus," February 10, 1862, David B. Harris Papers, DU; Polk to Jordan, March 18, 1862, *OR,* 7:438. The chain stretched across the river was swept away by a current that ran four miles per hour. Scheliha, *A treatise in coast defense,* 199.

28. Lynch to Polk, October 19, November 26, 1861, Montgomery Lynch CMSR; Wintter to Polk, November 12, 1861, D. Wintter, CMSR; Lynch to Polk, December 1, 1861, *OR,* 7:728–29. The *Memphis Appeal,* November 10, 1861, had a notice for several thousand Blacks to work at Fort Pillow and Island No. 10, noting that they should come with blankets and tools.

29. *St. Louis Republican* as quoted in *Memphis Appeal,* November 13, 1861.

30. Williams, *Beauregard,* 116–20; Roman, *Military Operations of General Beauregard,* 1:216; *Boston Journal* as quoted in *Richmond [Va.] Daily Dispatch,* August 26, 1863.

31. Page to Harris, March 3, 1861, and Corke to Harris, March 9, 1861, David B. Harris Papers, DU; Allardice, *Confederate Colonels,* 183; Polk, *Leonidas Polk,* 2:75–78; Nichols, *Confederate Engineers,* 52.

32. Nichols, *Confederate Engineers,* 77; Beauregard to Cooper, February 18, 1862, Mackall to Polk, December 10, 1861, and Polk to Mackall, January 17, 1862, *OR,* 7:890, 752, 837; Polk to Davis, March 11, 1862, *OR,* 10 (2): 311.

33. Johnston, *Narrative of Military Operations,* 85–86.

34. Report of Brigadier General George W. Cullom, March 4, 1862, Polk to Benjamin, March 2, 1862, and Polk to Jordan, March 18, 1862, *OR,* 7:437–38; Beauregard to Polk, February 23, 1862, and McCown to Polk, February 27, 1862, *OR,* 8:754, 760; Roman, *Military Operations of General Beauregard,* 1:361, 364; Daniel and Bock, *Island No. 10,* 45–67.

35. Sandidge to Ruggles, August 18, 1863, and De Saulles to Davis, September 8, 1861, Arthur B. De Saulles, CMSR; findagrave.com/memorial/147717525/arthur-brice-desaulles.

36. McCown to Polk, February 28, March 4, 5, 1862, *OR,* 7:762, 765, 766.

37. Report to Brigadier General J. Trudeau, Chief of Artillery, March 29, 1862, Harris to Pillow, April 1, 1862, and Mackall to Jordan, April 3, 1862, *OR,* 8:151–52, 806–7, 809; Wintter to McCown, March 26, 1862, D. Wintter CMSR.

38. Report of Captain D. B. Harris, March 9, 1862, *OR,* 8:139 ("unprecedented"); Harris to Jordan, March 9, 1862, David B. Harris Papers, DU.

39. Daniel and Bock, *Island No. 10,* 104–12.

40. Trautmann, ed., *A Prussian Observes,* 161; Daniel and Bock, *Island No. 10,* 113–14.

41. findagrave.com/memorial/90039440/victor-ernst_rudolf-wilhelm-von-scheliha, accessed March 30, 2020; Buckner to Cooper, May 25, 1863, Victor von Sheliha, CMSR; Gilmer to Mackall, March 9, 1862, Jeremy F. Gilmer, CMSR; Gilmer to wife, February 24, 1864, Jeremy F. Gilmer Papers, SHC.

42. Trautman, ed., *A Prussian Observes,* 160; Daniel and Bock, *Island No. 10,* 81–82.

43. Sheliha to Davidson, April 3, 1862, *OR,* 8:810–12; Sheliha, *A treatise on coast defense,* 29, 40; Victor von Shelhia, CMSR. In postwar years, Sheliha concurred with Beauregard and Joseph E. Johnston that the Confederates should have concentrated on "fewer points."

44. Roman, *Military Operations of General Beauregard,* 1:361.

45. Bragg to Jordan, March 18, 1862, Stewart to Jordan, March 21, 1862, abstract from monthly returns at Fort Pillow, April 30, 1862, and Villepique to Jordan, April 6, 1862, *OR,* 10 (2): 340, 352, 395, 476; Beauregard memorandum, March 3, 1862, *OR,* 7:915; Harris to Jordan, April 1, 1862, David B. Harris Papers, DU; Weinert, *The Confederate Regular Army,* 100.

46. Villepique to Jordan, April 6, 1862, *OR,* 10 (2): 395.

47. findagrave.com/memorial/153606352/andrew-belcher-gray, accessed March 17, 2020; Minor to Lockett, April 22, 1862, Minor Meriwether, CMSR.

48. Walke, "The Western Flotilla," 449.

2. Scandal at the Twin Rivers

1. Horn, *Leonidas Polk,* 185. For criticism of Harris, see Connelly, *Army of the Heartland,* 43–44; McMurry, *Two Great Rebel Armies,* 82–83; Smith, *Grant Invades Tennessee,* 6, 13.

2. Foster, "Building of Forts Henry and Donelson," 65; Creighton, *Life of Wilbur Fisk Foster,* 30, 32; Walke, "The Western Flotilla," 420; Wallace, "The Capture of Fort Donelson" 390 ("a happy gift"); *New York Times,* May 16, 1889 (Anderson); *Chicago Tribune,* February 20, 1862 (Dover description).

3. Warner, *Generals in Gray,* 74–75; Johnston, *Life of General Albert Sidney Johnston,* 407; Solonick, *Engineering Victory,* 226; Mitchell, *Tennessee State Gazetteer and Business Directory,* 218; John Haydon Journal, February [no dates], 1862, Benjamin F. Cooling Collection.

4. Johnston, *Life of General Albert Sidney Johnston,* 407; Cooling, *Forts Henry and Donelson,* 46.

5. Smith to Heiman, September 25, 1861, *OR,* 4:427–28. Even today Pine Bluff is heavily wooded and difficult to approach.

6. Cummings, *Yankee Quaker, Confederate General,* 75–76, 194, 176; Johnson to Harris, June 11, 1861, Isham Harris Papers, box 2, file 1, TSLA. Even Johnson's generally sympathetic biographer stated that the colonel's "first vital decision as a Confederate officer was an inept and tragic mistake." Richard McMurry (*The Fourth Battle of Winchester,* 71n8) concluded that the Confederates "had no choice but to put the installation [Henry] on very bad ground." By the time they gained access to Kentucky, they had invested time, energy, and other resources in constructing Fort Henry, and the general confusion and incompetence that were to characterize the Confederate high command in the West had set in. The rebel commanders concluded that they would be better off to continue work on Fort Henry."

7. Johnston, *Life of General Albert Sidney Johnston,* 410; US War Department, *The Official Military Atlas of the Civil War,* plate XI.3.

8. Creighton, *Life of Wilbur Fisk Foster,* 14; Wilbur F. Foster, CMSR.

9. Taylor, "The Defense of Fort Henry," 1:368–69.

10. Lindsley, ed., *Military Annals of Tennessee,* 879–80; findagrave.com/memorial/71032475/felix-r-r-smith; P. Tracy to Walker, March 24, 1861, H. T. Clary to Walker, April 18, 1861, in Thomas L. Estill, CMSR; 1860 US Census (ages for Smith and Mann); findagrave.com/

memorial/104048947/john-coleman-mann; Haydon Journal, February [no date], 1862, Benjamin F. Cooling Collection. For a photograph of Felix Smith, see findagrave.com/memorial/71032475/felix-randolph_robertson-smith.

11. Lindsley, ed., *Military Annals of Tennessee,* 879; Krick, *Staff Officers in Gray,* 79; *Columbia [Tenn.] Herald,* June 12, 1896.

12. Haydon Journal, February 1862, Benjamin F. Cooling Collection; Lindsley, ed., *Military Annals of Tennessee,* 879–80.

13. List of fortifications in the state of Tennessee, September 15, 1861, and Heiman to Polk, October 18, 1861, *OR,* 4:408, 460.

14. Polk to Mackall, November 10, 1861, *OR,* 3:310; Mackall to Dixon, September 30, 1861, and Polk to Mackall, October 7, 1861, *OR,* 4:433, 440; Orders No. 2, September 26, 1861, *OR,* 52 (2): 154; Polk to Johnston, November 28, 1861, *OR,* 7:710–11; Nichols, *Confederate Engineers,* 43; Connelly, *Army of the Heartland,* 80; "Capt. Jo Dixon," *Athens [Tenn.] Post,* February 21, August 29, 1862.

15. Warner, *Generals in Gray,* 105; Faust, ed., *Historical Times Illustrated Encyclopedia of the Civil War,* 311; Army, *Engineering Victory,* 92; Alexander to Walker, March 21, 1861, Jeremy F. Gilmer, CMSR; *Clarksville [Tenn.] Chronicle,* October 18, 1861; Johnston, *Life of General Albert Sidney Johnston,* 413; Connelly, *Army of the Heartland,* 80; Gilmer to wife, December 14, 1861, January 14, 1862, Gilmer Papers, SHC.

16. Gilmer to wife, October 15, 17, 19, 23, 1861, Gilmer Papers, SHC.

17. Gilmer to Mackall, November 3, 1861, Gilmer to Mackall, November 13, 1861, Johnston to Benjamin, November 15, 1861, and Gilmer to Mackall, November 3, 1861, *OR,* 4:506, 544, 554, 506; Gilmer to Mackall, November 26, 1861, *OR,* 52 (2): 221; *Clarksville [Tenn.] Jeffersonian,* August 9, 1861 (progress of bridge), as quoted in www.csa-railroads.com (website by David L. Bright); Gilmer to wife, October 25, 1861, Gilmer Papers, SHC; Connelly, *Army of the Heartland,* 72.

18. Sayers to Quarles, January 29, 1862, Sayers to Vest, August 9, 1862, Edward B. Sayers, CMSR; findagrave.com/memorial/140973004/edward-brydges-sayers; Gilmer to Mackall, November 3, 1861, *OR,* 4:506 (Morris).

19. Henry to Johnston, November 1, 1861, Gilmer to Mackall, November 3, 1861, and Johnston to Benjamin, November 8, 1861, *OR,* 4:497, 506, 528–29; Gilmer to Dixon, November 24, 1861, and Gilmer to Mackall, November 28, 1861, *OR,* 7:700, 710; Dixon to Mackall, October 4, 1861, *OR,* 52 (2) 167; Cooling, *Forts Henry and Donelson,* 56; Haydon Journal, February [undated], 1862, Benjamin F. Cooling Collection. William Preston Johnston (*Life of General Albert Sidney Johnston,* 414) concluded that the problems of the forts "were quite apparent" and "not the result of blind or careless policy," and the decision not to relocate was based on a "deliberate weighing of difficulties to advantages."

20. Henry to Johnston, November 1, 1861, Gilmer to Mackall, November 3, 1861, Gilmer to Mackall, November 4, 1861, and Henry to Johnston, November 7, 1861, *OR,* 4:496, 506, 514, 526.

21. US War Department, *The Official Military Atlas of the Civil War,* plates V.2, XI.5; Mahan, *A Treatise on Field Fortifications,* 19; "Dictionary of Fortification," lly.org/-rcw/cwf/dictionary/xgp-008.html, accessed August 19, 2020; Smith, *Grant Invades Tennessee,* 66.

22. Roman, *Military Operations of General Beauregard,* 1:216, 229; Johnston, *Life of Gen. Albert Sidney Johnston,* 409.

23. Gilmer to wife, November 4–December 14, 1861 (fourteen letters), Gilmer Papers, SHC; Haydon Journal, February [undated], 1862, Benjamin F. Cooling Collection.

24. Gilmer to Stevenson, November 26, 1861, and Gilmer to Sayers, December 10, 1861, EDLS.

25. Connelly, *Army of the Heartland,* 71–72. G. O. Watts, assistant engineer, also worked on the city's defenses.

26. Johnston, *Life of General Albert Sidney Johnston,* 416–17, 425; Haydon Journal, February [undated], 1862, Benjamin F. Cooling Collection; G. O. Watts to McKrew, December 6, 1861, Gilmer to Mackall, December 7, 1861, and Gilmer to Harris, December 11, 1861, *OR,* 739, 741, 745.

27. Mackall to Polk, October 28, 1861, *OR,* 4:481; Gilmer to Mackall, November 24, 28, 1861, *OR,* 7:698–700, 710.

28. Connelly, *Army of the Heartland,* 80–83, 85; Welsh, *Medical Histories,* 80.

29. Gilmer to Mackall, December 4, 1861, Gilmer to Dixon, December 4, 1861, and Pillow to Mackall, December 11, 1861, *OR,* 7:734–36, 758.

30. Gilmer to wife, December 20, 22, 29, 1861, January 2, 7, 1862, Gilmer Papers, SHC.

31. Roman, *Military Operations of General Beauregard,* 1:216; *New York Herald,* January 22, February 16, 1862; Gilmer to Mackall, December 21, 1861, Jeremy F. Gilmer, CMSR.

32. Haydon Journal, February [undated], 1862, Benjamin F. Cooling Collection; Johnston, *Life of General Albert Sidney Johnston,* 423–24; Report of Lieutenant Colonel Jeremy F. Gilmer, March 17, 1862, *OR,* 7:132. Army (*Engineering Victory,* 97) said that the failure to timely construct Fort Heiman fell not on Gilmer but Dixon. Yet, Tilghman, the officer in charge, never notified Johnston of the state of affairs, so the responsibility rests with him. Today Fort Heiman is a part of the Fort Donelson National Military Park.

33. Pickett, *Sketch of the Military Career of William J. Hardee,* 6.

34. Gilmer to wife, January 19, 23, 1862, Gilmer Papers, SHC; *Clarksville [Tenn.] Chronicle,* December 20, 1861 (500 slaves).

35. Report of Lieutenant Colonel Jeremy F. Gilmer, March 17, 1862, Report of Brigadier General Lloyd Tilghman, February 7, 1862, Report of Colonel A. Heiman, August 11, 1862, and Report of Lieutenant Colonel Jeremy Gilmer (second report), *OR,* 7:131–34, 137–46, 152, 131–32; Mahan, *A Treatise on Field Fortifications,* 19; Walter J. Morris, CMSR; Taylor, "The Defense of Fort Henry," 371 (waist-deep water); Johnston, *Life of General Albert Sidney Johnston,* 486.

36. *Memphis Appeal,* April 8, 1862; Report of Lieutenant Colonel Jeremy F. Gilmer, March 17, 1862, *OR,* 7:136 (Haydon captured); Smith, *Grant Invades Tennessee,* 117–18; Chernow, *Grant,* 172 ("harebrained scheme"); Taylor, "The Defense of Fort Henry," 371; *Chicago Tribune,* February 11, 1862.

37. Report of Brigadier General Lloyd Tilghman, February 7, 1862, *OR,* 7:139, 144; Johnston, *Life of General Albert Sidney Johnston,* 409.

38. Beauregard to Johnston, February 12, 1862, *OR,* 7:899.

39. Johnston to Benjamin, February 8, 1862, *OR,* 7:131; Roland, "Albert Sidney Johnston and the Defense of the Confederate West," 17; Cooling, *Fort Donelson's Legacy,* 12. In personal conversation, both Albert Castel and Steven Woodworth agreed with Cooling's conclusion.

40. Report of Brigadier General Simon B. Buckner, February 18, 1862, *OR,* 7:329; *National Tribune,* November 1, 1883; "River Batteries at Fort Donelson," 393–94; Smith, *Grant Invades Tennessee,* 210–11.

41. *Athens [Tenn.] Post,* August 29, 1862.

42. Report of Lieutenant Colonel Jeremy F. Gilmer, March 17, 1862, *OR,* 7:263; Report of Lieutenant Colonel Milton A. Haynes, March 22, 1862, *OR,* 7:146; "River Batteries at Fort Donelson," 393–94.

43. *Chicago Tribune,* February 19, 20, 1862. The odd design appears to have been an attempt at re-entrant angles, creating a crossfire.

3. Engineering in the Field

1. Hess, *Civil War in the West,* 39. The two sapper and miner outfits were Wintter's company, operating on the Mississippi River, and George Pickett's company, to be momentarily discussed, in Bowling Green.

2. Johnston, *Life of Gen. Albert Sidney Johnston,* 493.

3. CMSR of Miller, Rowley, Foster, Estill, Smith, Foster, Fremaux, Lynch, and Sayers.

4. Joseph A. Miller, CMSAR; Walter J. Morris, CMSR; Pickett, *Sketch of the Military Career of William J. Hardee,* 6; Henry N. Pharr, CMSR; findagrave.com/memorial/85768745/henry-h-pharr; findagrave.com/memorial/44009824/george-bibb-pickett, accessed June 21, 2020; "General R. B. Snowden and Staff," *Confederate Veteran* 3:186 (Helm); Rowland, *Encyclopedia of Mississippi History,* 1:333–35 (Helm biography); Scheliha, *A treatise on coast defense,* 190; findagrave.com/memorial/90015776/james-daniel-thomas; James D. Thomas, CMSR.

5. James Nocquet, CMSR; amboydepotmuseum.org/history.html, accessed April 15, 2020; *Chicago Tribune,* October 6, 1890.

6. Johnston, *Life of General Albert Sidney Johnston,* 495; vouchers, February 12, 16, 1862, George B. Pickett, CMSR; *New York Herald,* February 16, 25, March 1, 1862.

7. Gilmer to wife, February 22, 26, 27, 1862, Gilmer Papers, SHC; Smith, *Shiloh,* 25.

8. Gilmer to wife, March 3, 29, 1862, Gilmer Papers, SHC.

9. Ibid., March 9, 12, 24, 29, 1862; Fussell, "Narrative of Interesting Events," 384–95; Harris to Jordan, August 8, 1862, Harris Papers, SHC (Patterson); Sayers to Vest, August 9, 1862, Edward B. Sayer, CMSR.

10. Houghton, *Confederate Army of Tennessee,* 61; Johnston to Helm, March 18, 1862, *OR,* 10 (2): 338; Buell, "Operations in North Alabama," 702–3, 705.

11. Allardice, *Confederate Colonels,* 242; Kundahl, *Confederate Engineer,* 193; Lockett, "Surprise and Withdrawal from Shiloh," 604; findagrave.com/memorial/163898578/silvanus-wood-steele; Jason M. Fairbanks, CMSR; findagrave.com/memorial/41061272/jason-massey-fairbanks, accessed June 21, 2020.

12. Roman, *Military Operations of General Beauregard,* 1:245; Special Orders No. 44, February 29, 1862, Harris Papers, SHC; Geary, ed., *Celine,* xx; "Military Biography of Leon Joseph Fremaux," Leon J. Fremaux Papers, Civil War Collection, TU.

13. Kundahl, *Confederate Engineer,* xxi, 1, 11, 20, 22, 33, 50–52, 68, 69, 142.

14. US War Department, *The Official Military Atlas of the Civil War,* plate XII.5; Fremaux to General, April 6, 1862, Leon J. Fremaux, CMSR.

15. Special Orders No. 8, April 3, 1863, Headquarters, Second Corps, April 30, 1862, and Report of Lieutenant General Williams J. Hardee, *OR,* 10, pt. 1, 393, 469, 571; Special Orders No. 343, Minot Meriwether Biography, csa-railroads.com (website by David L. Blight); *Confederate Veteran* 3:187 (Helm); James Nocquet, CMSR. Wampler apparently did not arrive until after the battle.

16. Lockett, "Surprise and Withdrawal from Shiloh," 604–5; Pickett, *Sketch of the Military Career of William J. Hardee,* 10.

17. Lockett, "Surprise and Withdrawal from Shiloh," 605–6; Chisolm Report, April 14, 1862, SNBP; Welsh, *Medical Histories,* 79–80; Bragg to Beauregard, April 8, 1862, *OR,* 10 (2): 399; Gilmer to Beauregard, June 1, 1862, Jeremy Gilmer, CMSR.

18. Special Orders No. 20, 1862, *OR,* 10 (2): 415.

19. "Answers to interrogations," June 22, 1862, *OR,* 10 (1): 775; Smith, *Corinth 1862,* 15, 23; Roman, *Military Operations of General Beauregard,* 1:382; *New York Herald,* April 3, June 4, 5, 1862.

20. Harris to Lockett, May 8, 1862, Lockett Papers, SHC; Warner, *Generals in Gray,* 231–32; Lee to Beauregard, April 26, 1862, Special Orders No. 37, May 6, 1862, and Pegram to Beauregard, June 4, 1862, *OR,* 10 (2): 450, 500; 583 ("as the ground").

21. Fremaux map in Library of Congress Geography and Map Division, G3962.55551862cm; unsigned map in US War Department, *The Official Military Atlas of the Civil War,* plate XIV.2; "Answers to interrogations," June 22, 1862, *OR,* 10 (1): 775; "Biography of Minor Meriwether," in csa-railroads.com/essays/biography_of_minor_meriwether.htm, accessed April 4, 2020; George B. Pickett, CMSR; Nocquet to Gilmer, April 27, 1863, T. S. Newcomb, CMSR findagrave.com/memorial/184343020/george-blackwell-pickett, accessed April 14, 2020. Meriwether's sapper company would subsequently be sent to Vicksburg.

22. Kundahl, *Confederate Engineer,* 146, 150–51.

23. geni.com/people/Arthur-de-Saulles/6000000001944507435, accessed April 9, 2020; General Orders No. 1, May 20, 1862, *OR,* 10 (2): 534; Lucius D. Laudidy to Ruggles, August 1, 1863, Arthur B. de Saulles, CMSR; "Captain J. K. P. McFall," 656; James K. P. McFall, CMSR; Fergusson, "War Memories," William W. Fergusson Papers, TSLA ("jolly good fellow"); John G. Mann, CMSR.

24. findagrave.com/memorial/31871282/stephen-wilson-presstman, accessed April 9, 2020; *Alexandria [Va.] Gazette,* February 11, 1865; S. W. Presstman, CMSR; Kundahl, *Confederate Engineer,* 151; John W. Green, CMSR.

25. Lockett to wife, undated [June 1862?], Harris to Lockett, May 8, 1862, Lockett Papers, SHC; Walter J. Morris CMSR; Kundahl, *Confederate Engineers,* 155.

26. Sayers to Gilmer, August 3, 1863, Gilmer Papers, SHC (Fremaux and Pegram); http://www.csa-railroads.com/Essays/Biography/_of Minor_Meriwether.htm, accessed April 4, 2020; Geary, ed., *Celine,* 255n18; Kundahl, *Confederate Engineer,* 160; findagrave.com/memorial/85768745/henry-newton-pharr, accessed April 4, 2020; "Capt. H. N. Pharr's Death," *Forrest City [Ark.] Times,* October 29, 1897; John W. Green, CMSR.

27. McDonough, *War in Kentucky,* 43–47, 52–54; Hoffman, *"My Brave Mechanics,"* 88–89; General Reports, *OR,* 16 (1): 487.

28. McDonough, *War in Kentucky,* 54; Hoffman, *"My Brave Mechanics,"* 94; General Reports, *OR,* 16 (1): 392.

29. "Condition of Pickett's Company of Sappers and Miners," MC 28, Kennesaw National Battlefield Park; Nocquet to Gilmer, April 27, 1863, T. S. Newcomb, CMSR; Robertson, *River of Death,* 65; John Green to wife, November 8, 1862, John Green Letters, UVA; Sayers to Jack, February 1863, Edward B. Sayers, CMSR; Rush Van Leer, CMSR; 1860 US Census (Rush Van Leer); Conrad Meister, CMSR; Sayers to Gilmer, August 3, 1862, Gilmer Papers, SHC.

30. *Chattanooga [Tenn.] Daily Rebel,* November 1, 1862; Company D, 3rd Engineers, CMSR.

31. Boyd, *Military Reminiscences of Gen. William R. Boggs,* xi–xvii, 10, 13, 20, 66–67; emergingcivilwar.com/2018/06/09/William-freret-from-folly-to-war-to-success/. Both officers later followed Smith to the trans-Mississippi.

32. George R. Margraves, CMSR; findagrave.com/memorial/36771053/george-r-margraves.

33. Kundahl, *Confederate Engineer,* 161–63; Sayers to Jack, February 3, 1863, Edward B. Sayers, CMSR; Noe, *Perryville,* 64–66; Wampler to Bragg, September 6, 1862, Harris Papers, DU; Harris to Bragg, September 7, 1862, Bragg Papers, WRHS.

34. Harris to Bragg, September 15, 1862, "Memorandum of Water on road from Bardstown to Louisville via Shepherdsville," Harris to Sayers, September 6, 7, 1862, Harris to Hardee, September 9, 1862, all in Harris Papers, DU; Connelly, *Army of the Heartland,* 227–29.

35. Kundahl, *Confederate Engineer,* 168–69, 171–72; Pharr to Wampler, September 25, 1862, Bragg Papers, WRHS; Gilmer to Harris, October 7, 1862, David B. Harris Papers, DU.

36. Daniel, *Conquered,* 71–73; *OR,* 20 (2): 388, 393; Green to wife, November 8, 1862, John Green Letters, UVA; H. N. Pharr, CMSR.

37. Gilmer to Harris, October 7, 1862, David B. Harris Papers, DU; Gilmer to Breckinridge, October 7, 11, 1862, James Nocquet CMSR; Green to wife, November 8, 1862, John Green Letters, UVA.

38. Robert P. Rowley, CMSR; Henry C. Force, CMSR; Gilmer to Bragg, December 7, 1862, *OR,* 20 (2): 443; Gilmer to Cooper, November 10, 1862, Edward B. Sayers CMSR.

39. Nocquet to Breckinridge, November 11, 1862, *OR,* 20 (2): 398–99.

40. Gilmer to Bragg, December 7, 1862, Jeremy Gilmer, CMSR; James Nocquet, CMSR; Kundahl, *Confederate Engineer,* 185, 187; Robertson, *River of Death,* 65; Allardice, *Confederate Colonels,* 183.

41. John C. Wrenshall, CMSR; findagrave.com/memorial/82267471/john-c_-wrenshall; Kundahl, *Confederate Engineer,* 226–27.

42. Kundahl, *Confederate Engineer,* 188, 190–91; "With Cumberland University 49 Years," *Confederate Veteran,* 421; James K. P. McFall, CMSR.

43. Shiman, "Engineering and Command," 90–92.

44. Daniel, *Battle of Stones River,* 24–26, 29–30; Walker to Flynt, December 29, 1862, and Martin to Davis, January 25, 1863, *OR,* 20 (2): 272, 362.

45. Daniel, *Battle of Stones River,* 198–99.

46. Steele to Polk, January 5, 1863, *OR,* 20 (2): 485.

4. Confronting Challenges

1. Daniel, *Conquered,* 215; Hazlett et al., *Field Artillery Weapons of the Civil War,* 107.

2. Hess, *The Knoxville Campaign,* 2–3.

3. Reports of Major General Horatio G. Wright, *OR,* 20 (1): 86, 90, 91; Report of Charles J. Walker, *OR,* 20 (1): 94; Wright to Cullom, December 18, 1862, Marshall to Jones, and Seddon to Smith, January 4, 1862, *OR,* 20 (2): 199–200, 472, 483; *Richmond [Va.] Daily Dispatch,* January 2, 3, 1863.

4. Seddon to Smith, January 4, 1863, and Gilmer to Grant, January 5, 1863, *OR,* 20 (2): 484, 486.

5. Grant to Gilmer, January 9, 11, February 27, 1863, Grant to Rives, February 6, 1863, Grant to Haydon, February 1, 1863, and Grant to Maxwell, February 23, 1863, Lemuel P. Grant Papers, AHS; Foster to Martin, April 5, 1863, *OR,* 23 (2): 742; *Richmond [Va.] Daily Dispatch,* March 27, 1863; Gilmer to Haydon, January 6, 1863, Gilmer to Robinson, March 27, 1863, Gilmer to Norquet, July 20, 1863, A. L. Maxwell to Confederate States, September 30, 1863, csa-railroads.com (website by David L. Bright).

6. Richard C. McCalla Biographical Sketch, AU; findagrave.com/memorial/68711552/richard-calvin-mccalla; Mitchell, *Tennessee State Gazetteer and Business Directory,* 71; *Memphis Public Ledger,* February 9, 1871; *Memphis Appeal,* February 18, 1871; findagrave.com/memorial/23926213/waightsill-avery-ramsey; Waightsill Ramsey, CMSR; findagrave.com/memorial/129171583/matthew-fontaine-maury; "Matthew F. Maury," *Confederate Veteran,* 29.

7. Daniel, *Conquered,* 135, 143–45; American Society of Civil Engineers, *Transactions of the American Society of Civil Engineers,* vol. 9, 545; Cobb to Machin, April 13, 1863, Robert L. Cobb, CMSR.

8. Hardee to Bragg, January 26, 1863, *OR,* 23 (2): 617; Northrop to Davis, February 1, 1864, *OR,* 32 (2): 674; Gilmer dispatches, February 11, 19, 1863, Tennessee & Alabama Railroad, csa-railroads.com (website by David L. Bright).

9. Brent to Presstman, February 3, 1863, S. W. Presstman, CMSR; Wampler to Gilmer, June 27, 1863, Gilmer to Wampler, July 3, 1863, and Special Orders No. 34, June 7, 1863, J. W. Wampler, CMSR.

10. Kundahl, *Confederate Engineer,* 203, 208–16; George M. Helm, CMSR.

11. Johnston to Davis, April 15, 1863, Sayers to Jack, June 12, 29, 1863, *OR,* 23 (2): 760, 874, 891; H. D. Clayton to wife, May 23, 1863, www.tennesseecivilwarsourcebook.com, accessed April 2, 1863. Extensive works were also built around Shelbyville, which were described by a *New York Herald* reporter on July 8, 1863, as "the work of a skilled engineer."

12. Ruger to Pittman, October 15, 1863, and Knipe to Pittman, *OR,* 30 (4): 399–400, 427.

13. Kundahl, *Confederate Engineer,* 211–13.

14. "List of Officers and Assistants, Engineer Corps, Army of Tennessee, March 16, 1863," Bragg Papers, WRHS.

15. Various vouchers in S. W. Presstman, CMSR; Gilmer to Presstman, June 1, July 22, 25, 1863, Letters and Telegrams Sent, CED, Engineer Department, 1861–1864, RG 109, National Archives; J. K. P. McFall, CMSR; Sayers to Jack, June 29, 1863, *OR,* 23 (2): 891; Sayers to Jack, July 24, 1863, Edward B. Sayers, CMSR.

16. Ramage, *Rebel Raider,* 160, 181; Amos S. Darrow, CMSR.

17. Steele to Rives, June 2, 1863, Steele to Gilmer, July 8, 1863, John F. Steele, CMSR; Gilmer to Steele, June 8, 1863, Letters and Telegrams Sent, CED.

18. Presstman to Gilmer, February 14, 1863, Green to Gilmer, February 15, 1863, George W. Buchanan to Jones, February 27, 1863, Gilmer to Jones, March 3, 1863, H. R. Pharr, CMSR; Hardee to Cooper, November 6, 1862, Hardee to Gilmer, June 5, 1863, Helm to Hardee, June 5, 1863, Wampler to Gilmer, July 6, 1863, George M. Helm, CMSR; "List of Nominations for Appointment," May 20, 1863, Jeremy F. Gilmer, CMSR. The *New York Herald* (July 8, 1863) somehow got the names of four engineers—S. W. Steele, H. C. Forie [Force], H. [A] H. Buchanan, and L. P. R. [J. K. P.] McFall.

19. Buckner to Cooper, May 25, 1863, Victor Von Sheliha, CMSR; "Return of Officers and Men Hired," June 1863, James Norquet, CMSR; Kundahl, *Confederate Engineer,* 219–22, 229, 231–33, 250.

20. Johnston, *Narrative of Military Operations,* 190; abstract from field return of Army of Mississippi and Eastern Louisiana, July 30, 1863, *OR,* 24 (3): 1039.

21. James M. Couper, John G. Kelly, S. H. Lockett, CMSR; Geary, ed., *Celine,* 113.

22. Hughes, *General William J. Hardee,* 158–61; General Orders No. 2, July 25, 1863, *OR,* 24 (3) 1031; memorandum for Lieutenant General Hardee, August 31, 1863, *OR,* 30 (4): 573.

23. Gilmer to Grant, August 13, 1863, Gilmer to Johnston, August 14, 18, 1863, *OR,* 30 (4): 493–94, 496, 503–4.

24. Robinson to Ginder and Vinet, November 27, 1863, *OR,* 31 (3): 753–54; findagrave.com/memorial/138291670/john-baptists-vinet.

25. McDonough, *William Tecumseh Sherman,* 452–53; Polk to Loring, February 12–13, 1864, *OR,* 32 (2): 724, 732. Jones was probably William A. C. Jones, born in Kentucky in 1829. See findagrave.com/memorial/39730432/william-a-c-jones; Polk to Loring, February 13, 1864, *OR,* 32 (2): 732; findagrave.com/memorial/31934734/richard-a.-o'hea.

26. georgiaencyclopedia.org/articles/history-archeology/lemuel-grant-1817-1893; *The Biographical Directory of the Railroad Officials of America for 1887,* 130; Davis, *What the Yankees Did to Us,* 17.

27. Grant to Thomasey, December 14, 1862, January 14, 1863, Grant to Gilmer, December 25, 1862, Grant to Dexter, February 13, 1863, Rives to Grant, September 9, 1863, Lemuel P. Grant Papers, AHS; findagrave.com/memorial/111254490/amory-grant.

28. Rives to Grant, October 16, 23, 1862, Gilmer to Grant, October 22, 1862, March 4, May 5, 23, 1863, Gilmer to Grant, November 1, 13, 1862, January 5, May 4, June 5, 1863, and Grant to Green, November 13, 17, December 16, 31, 1862, Lemuel Grant Papers, AHS; Randolph to Grant, October 22, 1862, *OR,* Series 4, 2:139; Rives to Grant, September 16, 1862, and Gilmer to Green, November 17, 1862, csa-railroads.com (website by David Bright).

29. Gilmer to Grant, January 17, 1863, Grant to Rives, April 4, 1864, and Grant to Presstman, May 12, 1864, Lemuel Grant Papers, AHS. See also Rives to Grant, March 28, April 28, 1864, and Grant to Presstman, May 7, 1864, Letters and Telegrams Sent, CED.

30. Gilmer to Grant, June 23, July 16, 1863, Lemuel Grant Papers, AHS.

31. Davis, *What the Yankees Did to Us,* 59–60; Gilmer to Grant, August 11, 1863, *OR,* 30 (4): 489; Gilmer to Wright, October 21, 1863, *OR,* 31 (3): 575; Grant to Rowley, November 4, 1863, Lemuel Grant Papers, AHS; Fryman, "Fortifying the Landscape," 50, 54; Hess, *Fighting for Atlanta,* 150–54, 158–61.

32. William A. Hansell, CMSR; findagrave.com/memorial/30457216/william-andrew-hansell. Theodore Moreno was born in Florida and attended the University of Virginia, where he took classes in practical engineering and worked for different railroads. See Theodore Moreno, CMSR.

5. Engineering Colossus

1. Beauregard to Cooper, September 24, 1862, *OR,* 15:810–13; Beauregard to Harris, April 21, 1862, David B. Harris Papers, DU.

2. Smith, *The Union Assaults at Vicksburg,* 14.

3. *Memphis Appeal,* August 7, 1862.

4. Report of M. L. Smith, *OR,* 15:6; Faust, ed., *Historical Times Illustrated Encyclopedia of the Civil War,* 697. In July 1866, at the age of forty-seven, Smith died in Athens, Georgia, after a brief illness.

5. Lockett, "The Defense of Vicksburg," 483.

6. *Chicago Tribune,* July 3, 19, 1862.

7. Scheliha, *A treatise on coast defense,* 30–31; Shea and Winschel, *Vicksburg Is the Key,* 22.

8. Shea and Winschel, *Vicksburg Is the Key,* 98–99.

9. Johnston, *Narrative of Military Operations,* 152; Hogane, "Reminiscences of the Siege of Vicksburg," 224–25.

10. Bearss and Grabau, "How Porter's Flotilla Ran the Gauntlet Past Vicksburg," 48.

11. Solonick, *Engineering Victory,* 60; Grabau, *Ninety-Eight Days,* 20–23, 43–44.

12. Grabau, *Ninety-Eight Days,* 41–43; *New York Herald,* July 13, 1862; *Chicago Tribune,* July 19, 1862; Higgins to Blakely, February 12, 1863, *OR,* 24 (3): 623.

13. Shea and Winschel, *Vicksburg Is the Key,* 103–3.

14. Harris to Lockett, May 8, 1862, Lockett Papers, box 7, SHC; Lockett, "The Defense of Vicksburg," 483, Lockett to wife, August 4, 1862, Lockett Papers, SHC.

15. Lockett to wife, September 22, November 5, 25, December 17, 1862, Lockett Papers, folder 8, SHC.

16. Lockett, "The Defense of Vicksburg," 483; Lockett to Gilmer, February 13, 1863, Samuel H. Lockett, CMSR.

17. I. J. Thysseus, CMSR.

18. Mahan recommendation, November 15, 1856, and Robinson to Seddon, April 12, 1863, James M. Couper, CMSR.

19. Lockett to Rives, December 1, 1863, George Donnellan, CMSR; www.gilderlehrman.org/collection/glc09263; Report of Major Samuel H. Lockett, *OR,* 24 (2): 330; findagrave.com/memorial/18750827/george-donnellan.

20. Wharton to President, January 18, 1861, Powhatton Robinson, CMSR; Lockett, "The Defense of Vicksburg," 484–85; US War Department, *The Official Military Atlas of the Civil War,* plate XXXVII.1. Some of the remains of Fort Pemberton survive.

21. R. R. Southard, CSMR; J. J. Conway, CMSR; publications.newberry/hogane-and-lambach-map-city-davenport-and-its-suburbs-1857, accessed October 11, 2020; Henry Ginder Biography, Henry Ginder Papers, Louisiana Research Center, TU.

22. Report of Major Samuel H. Lockett, *OR,* 24 (2): 335; digitalcollections.nypl.org/items/510d47e0-1414; territorialkansasonline.ku/index.php?SCREEN; pps-west.com/product/very-early-view-of-denver, all accessed October 23, 2020.

23. dc.lib.unc.edu/cdm/singleitem/collection/gilmer/id/96/rec/1.

24. "Confederate Topographical Report of Port Hudson," August 30, 1862, mss 3921, LSU; Geary, ed., *Celine,* 255n18.

25. Gillimard to Lambert, January 3, 1863, J. V. Gillimard, CMSR; *Gardner's New Orleans Directory for 1861,* 184; Weinert, *The Confederate Regular Army,* 104–6.

26. "John G. Kelly," 370; Pettus to Reeve, April 8, 1863, Kelly to Pettus, April 8, 1863, Kelly to Pettus, April 16, 1863, and Pettus to Reeve, April 18, 1863, *OR,* 24 (3): 725, 727–28, 749–51, 763–64; S. R. Tresilian report, *OR,* 24 (2): 208.

27. Solonick, *Engineering Victory,* 225–29. The three Confederate engineers who previously served in the US Corps of Engineers were M. L. Smith, David Harris, and S. H. Lockett. Harris, of course, was not present during the siege. Solonick states that his list is a "rough estimate," but "is by no means definitive."

28. Johnston, "Jefferson Davis and the Mississippi Campaign," 474.

29. Lockett, "The Defense of Vicksburg," 483–84; reports of Captain Frederick E. Prime and Cyrus B. Comstock, *OR,* 24 (2): 169–70; report of Major Samuel H. Lockett, *OR,* 24 (2): 330.

30. Hess, *Storming Vicksburg,* 15–18; Field, *American Civil War Fortifications,* 34–44; Grabau, *Ninety-Eight Days,* 45–47.

31. Winschel to Daniel, October 15, 2020. Timothy Smith concurred with Winschel. "The issue of turning the line at the Hall's Ferry Road was the Valley of Stout's Bayou. Obviously, it had to be crossed somewhere, but doing it farther south put the line crossing the valley at a place where it was much narrower with disjointed ridges to cover angles." Smith to Daniel, October 16, 2020. Unfortunately, the sector south of the Salient Work is not now a part of the Vicksburg National Military Park, so it is difficult, if not impossible, to get a concept of the 1863 topography. A 1918 map of Vicksburg reveals that even at that early date the city environs had extended well to the south toward Warrenton.

32. Johnston, "Jefferson Davis and the Mississippi Campaign," 474; Johnston, *Narrative of Military Operations,* 152.

33. Ballard, *Pemberton,* 147–49.

34. Taylor to Forney, May 13, 1863, and Lockett to Memminger, February 10, 1863, *OR,* 24 (3): 872, 621.

35. De Veure to Beauregard, February 18, 1863, Henry De Vuere, CMSR; findagrave.com/memorial/84187911/henry-de_-veure; De Veure to Memminger, May 9, 1863, *OR,* 23 (3): 848; Winschel to Daniel, October 30, 2020.

36. Winschel, *Triumph and Defeat,* 96; Johnston, *Narrative of Military Operations,* 209. A. S. Abrams, a young Mississippi artilleryman, described the west bank of the Big Black as "an almost precipitate height, overlooking the east shore, and forming a succession of lofty cliffs." Abrams, "A Full and Detailed History of the Siege of Vicksburg," 77.

37. Lockett, "The Defense of Vicksburg," 488; Hogane, "Reminiscences," 291–93.

38. Hess, *Storming Vicksburg,* 49; Smith, *The Union Assaults at Vicksburg,* 123–26.

39. Grabau, *Ninety-Eight Days,* 409.

40. Report of Brigadier General F. A. Shoup, report of Ashbel Smith, reports of Brigadier General John C. Moore, *OR,* 24 (2): 407, 386–87, 380, 905; Forney to Chief of Staff, May 21, 1863, *OR,* 21 (3): 905.

41. *Chicago Tribune,* May 29, June 2, 1863; Smith, *The Union Assaults at Vicksburg,* 346; Hess, *Storming Vicksburg,* 290–92, 295.

42. Report of Major Samuel H. Lockett, *OR,* 24 (2): 331; Bearss, *The Campaign for Vicksburg,* 3:886–87.

43. Solonick, *Engineering Victory,* 80, 82, 181–82; *Chicago Tribune,* June 10, 30, 1863.

44. Sims, ed., "A Louisiana Engineer at the Siege of Vicksburg," 374, 376; Major Samuel H. Lockett report, *OR,* 24 (2): 332; *Montgomery Daily Advertiser,* August 5, 1863.

45. Grabau, *Ninety-Eight Days,* 428–438; Lockett, "The Defense of Vicksburg," 491–92; Solonick, *Engineering Victory,* 181–83, 187–203.

46. Report of Major Samuel H. Lockett, *OR,* 24 (2): 334.

47. *Chicago Tribune,* July 14, 1863.

48. fampeople.com/cat-alexander-st-clair-abrams_2; Abrams, "A Full and Detailed History of the Siege of Vicksburg," 10, 14–15, 44, 66–67. Portions of the Abrams pamphlet also appeared in the *Mobile Advertiser and Register,* July 19, 1863.

49. Report of Captain Frederick E. Prime, *OR,* 24 (2): 170, 176; Solonick, *Engineering Victory,* 202, 213–14. An example of this Yankee ingenuity was the 120-foot-long bridge built over the Big Black River within six hours—not of pontoons, but of buoyant cotton bales. See *OR,* 23 (2): 203.

6. Organizing Engineer Troops

1. Army, *Engineering Victory*, 5, 6, 22.

2. Ibid., 22; US Census, 1860, Davidson County, Tennessee; Hoffman, *"My Brave Mechanics,"* 14–17.

3. Swint and Mohler, "Eugene F. Falconnet," 219–22; findagrave.com/memorial/109720590-eugene-frederic-falconnet; Cathey and Wadday, *"Forward My Brave Boys!,"* 314; *Williams' Memphis City Directory for 1860*, 357. Civil engineers in Memphis who never served in the army or under contract included James D. Cook, Charles F. Johnson, John N. McNalty, Joseph McWilliams, J. F. Parsons, John B. Young, and P. H. Hammarskald. I have been able to trace at least fifteen other confirmed Tennessee civil engineers who had no connection with the Confederate army, either through enlistment or contract. See Mitchell, *Tennessee State Gazetteer and Business Directory*, 375.

4. "Classified Population of the States and Territories," 1860, www.2.census.gov/library/publications/decennial/population/1860a-04.pdf. Railroad civil engineers compiled from csa-railroads.com (website by David Bright).

5. Hess, *Field Armies and Fortifications*, 18–21; Gilmer to Seddon, December 24, 1862, *OR*, Series 4, 2:259–60.

6. Gilmer to wife, October 12, 16, 1862, Gilmer Papers, SHC.

7. Lash, *Yankee Bridge Builder*, 112–16; *Memphis Appeal*, September 2, 1862; *Chattanooga [Tenn.] Daily Rebel*, September 16, 26, October 15, 1862; Glover entry, October 15, 1862, Jones to Myers, October 20, November 6, 1862, Gilmer to Randolph, November 6, 1862, Memphis & Charleston File, csa-railroads.com, and Stringfellow to Jones, September 20, 1862, Jones to Gorgas, September 28, 1862, Stringfellow to Maxwell, October 2, 1862, and Rives to Smith, November 5, 1862, Nashville & Chattanooga File, csa-railroads.com (website by David L. Bright).

8. Gilmer to Seddon, December 26, 1862, *OR*, Series 4, 2:259–61. The four sapper companies were those of Wintter, Gillimard, Pickett, and Winston.

9. Jackson, *First Regiment Engineer Troops*, 2; "An Act to provide and organize engineer troops," March 6, 1863, and General Orders No. 104, July 23, 1863, *OR*, Series 4, 2:445–46, 678–79; General Orders No. 66, May 22, 1863, *OR*, 25 (2): 817–18; Seddon to Lee, July 25, 1863, *OR*, 27 (3): 1038.

10. Crute, *Units of the Confederate States Army*, 65–67; Talcott, "Reminiscences of the Confederate Engineer Service," 258; Blackford, *War Years with Jeb Stuart*, 251; Jackson, *First Regiment Engineer Troops*, 15–30.

11. Third Engineers, CSMR; Arthur W. Gloster, CMSR.

12. James R. Davies, CMSR; William D. Prinz, CMSR; Menefee Huston, CMSR; T. J. Ridly, CMSR; 1860 US Census, Tennessee; findagrave.com/memorial/67914396/menefee-huston. Lieutenant Henry Otey Minor (1839–1928) is found on findagrave.com/memorial/135602541/henry-minor, but his occupation is not listed.

13. Sayers to Gilmer, July 23, 1863, Sayers to "Sir," April 14, 1864, Edward B. Sayers, CMSR; Rives to Presstman, September 9, 1863, Letters and Telegrams Sent, CED.

14. No. 40 Special Requisition, August 1, 1863, Robert L. Cobb, CMSR.

15. Notes to Lieutenant W. B. Richmond, June 29, 1863, and report of Brigadier General Bushrod R. Johnson, *OR*, 23 (1): 622, 608; White and Runion, eds., *Great Things Are Expected of Us*, 99; Daniel, *Conquered*, 162–71.

16. Sayers to Jack, February 7, 1863, and voucher, July 31, 1863, in Edward B. Sayers, CMSR; Presstman to Rives, November 8, 1863, T. S. Newcomb, CMSR; Polk to Mackall, July 5, 1863, *OR,* 23 (1): 626; Jack to Hardee, July 4, 1863, Hardee to Jackson, July 4, 1863, Pickett to Polk, July 5, 1863, and abstract from return of the troops, July 31, 1863, *OR,* 23 (2): 898–900, 941.

17. Krick, *Staff Officers in Gray,* 322; Special Orders No. 39, August 3, 1863, *OR,* 23 (2): 949; Green to Adjutant General, November 28, 1861, and Clingman to Secretary of War, December 10, 1861, Thaddeus C. Coleman, CMSR; Francois I. J. Thysseus, CMSR.

18. Wilder to Garfield, August 22, 1863, *OR,* 30 (3): 122; Robertson, *River of Death,* 283–89; Hoffman, *"My Brave Mechanics,"* 162–63.

19. Daniel L. Kelly to Miss Honnell, August 28, 1863, Daniel L. Kelly Letters, EU.

20. Companies A and C, 3rd Confederate Engineers, CMSR.

21. Fergusson, "War Memories," William W. Fergusson Papers, TSLA.

22. Report of Major General Thomas C. Hindman, *OR,* 30 (2): 293; Powell, *The Chickamauga Campaign,* 1:143–49.

23. Sayers to Jack, September 24, 1863, Edward B. Sayers, CMSR; report of Colonel William Grose, *OR,* 30 (1): 782; findagrave.com/memorial/140973004/Edward-bryudges-sayers; McCalla to wife, October 4, 1863, Richard McCalla Letters, AU; Fergusson, "War Memories," William W. Fergusson Papers, TSLA.

24. Gilmer to Cooper, October 7, 1864, James Norquet, CMSR; *Chicago Tribune,* October 6, 1890; findagrave.com/memorial/195978416/emelia-nocquet. Kundahl (*Confederate Engineer,* 266) suggests that Noquet may have been assisted in his desertion by Captain Francois Thysseus, but there is no evidence that they had ever met. The *Richmond Enquirer* (December 3, 1863) noted that Bragg's chief engineer "basely deserted to the enemy," although it incorrectly named him as James Hallonquist, Bragg's chief of artillery.

25. Companies B, D, F, 3rd Confederate Engineers, CMSR; Presstman to Ramsey, October 15, 1863, horsesoldiers.com/images/product/17/56503.jpg; McCalla to wife, October 19, 30, November 9, 1863, Richard McCalla Letters, AU.

26. Faust, ed., *Historical Times Illustrated Encyclopedia of the Civil War,* 427; Danville Leadbetter, CMSR; Gallagher, ed., *Fighting for the Confederacy,* 323–24; Bragg to Longstreet, November 22, 1863, *OR,* 31 (3): 736, 780; Welsh, *Medical Histories,* 131; *New York Herald,* September 29, 1866. Colonel William W. Blackford, commanding the 1st Engineers, later mused that any engineer over forty in the Old Army proved worthless. They felt that they had nothing to learn, saw others as inferiors, and they were "narrow-minded, selfish, and bigoted," having been "fossilized" by garrison life.

27. Shakelford dispatch, October 15, 1863, *OR,* 30 (2): 595; Shakelford to Richmond, October 16, 1863, Shakelford to Burnside, October 17, 1863, Jones to Seddon, October 17, 1863, *OR,* 30 (4): 450, 595, 762; *Memphis Appeal,* November 7, 1863, *Richmond Dispatch,* October 19, 1863, and January 23, 1864, Rives to Longstreet, November 23, 1863, Bleyes to Robinson, December 9, 1863, all in csa-railroads.com (website by David Bright).

28. McCalla to wife, January 8 and February 4, 1864, Richard McCalla Letters, AU. The equipment in Company A included: 1 bellows, 1 anvil, 1 vice, tongs, hand hammers, sledge hammers, 1 screw plate and screws, files, cast iron, nippers, 4 log chains, 4 fifth chains, 4 wagon cloths, 4 dozen axes, 3 dozen picks, 30 shovels, 2 frows, 1 grindstone, whetstones, 2 tapelines, 6

2-foot rules, 2 jack planes, 2 jointing planes, 2 force planes, 2 smoothing planes, 100 horseshoes, 100 mule shoes, horseshoe nails, rope, block and tackle, 6 hatchets, 1 bevel, 2 try squares. See Partin, "The Civil War in East Tennessee," 241n12.

29. Hess, *The Knoxville Campaign,* 30–32; Daniel, *Conquered,* 232–33. Bragg wrote that he wished to "get rid of him [Longstreet] and see what he could do on his own resources."

30. Report of Lieutenant General James Longstreet, *OR,* 31 (1): 456; Longstreet, *From Manassas to Appomattox,* 486.

31. Gallagher, ed., *Fighting for the Confederacy,* 313; Allardice, *Confederate Colonels,* 101; Hess, *The Knoxville Campaign,* 127; report of Lieutenant General James Longstreet, *OR,* 31 (1): 460; findagrave.com/memorial/6050771/thomas-jefferson-moncure.

32. McCalla to wife, October 30, November 9, 20, 1863, Richard McCalla Letters, AU. The companies were still at Charleston on November 20, McCalla adding that they "now have but little to do."

33. Hess, *The Knoxville Campaign,* 37–38; Gallagher, ed., *Fighting for the Confederacy,* 313, 315; report of Lieutenant General James Longstreet, *OR,* 31 (1): 456; Wright to Reeves, October 27, 1863, and Longstreet to Bragg, *OR,* 31 (3): 596–97, 707; Company B, 3rd Engineers file, CMSR; Sorrell, *Reflections of a Confederate Staff Officer,* 206.

34. Hess, *The Knoxville Campaign,* 149, 155, 157, 170–71.

35. Gallagher, ed., *Fighting for the Confederacy,* 323, 325, 327.

36. "Report of Brigadier Gen. James Patton Anderson," in Hoffman, ed., *The Confederate Collapse at the Battle of Missionary Ridge,* 34; Tower, ed., *A Carolinian Goes to War,* 134–35.

37. *Augusta [Ga.] Weekly Constitutionalist,* November 27, 1863; Company D and Company G, 3rd Engineers, CMSR; Walter J. Morris, CMSR; abstract from return, December 31, 1863, *OR,* 31 (3): 883. It is possible that the separated troops are counted in the December 31 return as present, but were not actually at Dalton.

38. Daniel, *Conquered,* 251; Special Orders No. 4, January 12, 1864, *OR,* 32 (2): 549.

39. Company G, 3rd Engineers, CMSR; Pickett to Hardee, December 5, 1863, Doug Schanz Collection.

40. Wynne and Taylor, eds., *This War So Horrible,* 34, 36, 40, 49; Jenkins, "Saving Dalton's Battlefields," 19. Captain John R. Oliver is thought to be John Robert Oliver, born 1837. Findagrave.com/memorial/17978094/john-robert-oliver.

41. Companies C, D, G, all in 3rd Engineers, CMSR; Gibbons to Lawton, *OR,* 32 (3): 772; Johnston to Davis, January 15, 1863, *OR,* 32 (2): 559; Glenn to Ewell, August 9, 1863, and A. P. Stewart to Hood, April 1, 1864, John W. Glenn, CMSR; Presstman to Grant, April 25, 1864, and Rives to Grant, January 18, 1864, Lemuel P. Grant Papers, AHS; Skellie, ed., *Lest We Forget,* 2:644.

42. General Orders No. 17 and General Orders No. 21, *OR,* 32 (2): 631, 743; Tower, ed., *A Carolinian Goes to War,* 164; Orlando M. Poe reports, *OR,* 44:59 (350 entrenching tools); Bragg, *Joe Brown's Army,* 71–72; "Worthy Widow Who Deserves a Pension," 318; findagrave.com/memorial/15951041/t.-george-raven.

43. Hess, *Fighting for Atlanta,* 20; Special Order No. 106, May 6, 1864, *OR,* 38 (4): 671; George H. Hazlehurst, D. W. Currie, G. H. Browne, and Thomas E. Marble, CMSR; hazlehurstga.gov/ColonelHazlehurst.aspx, accessed December 21, 2020; findagrave.com/memorial/61584900/Thomas-elan-marble.

7. The Mapmakers

1. Beers, "A History of the US Topographical Engineers, 1813–1863," Part 1, 287, 289, Part 2, 348–49; Schubert, ed., *The Nation Builders,* 8–9, 58–59, 74. The topographical engineers had an office in Knoxville prior to the Civil War and mapped segments of the Tennessee River, but I have been unable to discern precisely what maps they produced and had on file.

2. Smith to Heiman, September 25, 1861, *OR,* 3:427; Campbell, "The Lost War Maps of the Confederates," 480.

3. teva.content.oclc.org/customizations/global/. . . . , accessed March 19, 2020.

4. For a copy of the Fremaux map of roads from Corinth to Pittsburg Landing, see US War Department, *The Official Military Atlas of the Civil War,* plate XII.5; Corinth Fortifications, No. GLC05668, Gilder Lehrman Collection, Chestertown, Maryland; Headquarters Polk's Corps, February 28, 1863, *OR,* 20 (1): 692; US War Department, *The Official Military Atlas of the Civil War,* plate XXXI.1.

5. US War Department, *The Official Military Atlas of the Civil War,* plates XXXVII.1, XLVI.1, XLVI.2, XLVI.4, XLVII.2, XLVII.3, XLVII.7; geographics.com/P/RareMaps/rugerededward, accessed September 10, 2020.

6. detecting.us/wp-content/uploads/2018/11/Perryville-Ky-high-resolution-1862.jpg, accessed April 15, 2020; map of Murfreesboro, December 29, 1862, No. GLC05668, Gilder Lehrman Collection; US War Department, *The Official Military Atlas of the Civil War,* plate XXXI.2; vouchers, February 28, June 15, 1863, Conrad Meister, CMSR; findagrave.com/memorial/98915685/conrad-meister.

7. Sheridan to Garfield, July 25, 28, 1863, *OR,* 23 (2): 556–57, 564.

8. *New York Herald,* August 16, 1863. Lieutenant Fergusson would later recall: "Meister, who had been with us at Tupelo, Mississippi, pretended that he would like to take up field work and obtaining copies of all our maps went off and never returned." Fergusson, "War Memories," William W. Fergusson Papers, TSLA.

9. Gilmer to Presstman, July 22, 25, 1863, Letters and Telegrams Sent, CED; Benjamin F. Cheatham Civil War map of area southeast of Murfreesboro, Tennessee, TSLA; E. G. Anstey, CMSR; US War Department, *The Official Military Atlas of the Civil War,* plate XLVIII.1.

10. https://www.chattanoogan.com/2008/3/11/123750/Hamilton-County-Pioneers---the-Tyners.aspx; John S. Tyner, CMSR; Tyner and Steele maps, John Wheeler Papers, ADAH; Hughes, *The Pride of the Confederate Artillery,* 313; US War Department, *The Official Military Atlas of the Civil War,* plates XXXI.5, XXXII.5.

11. Foster, "Battlefield Maps in Georgia," 369–70; Fergusson, "War Memories," William W. Fergusson Papers, TSLA. For copies of the Clayton maps see Henry Clayton Papers, ADAH.

12. Buchanan, "From Engineer for the Army of Tennessee," 370.

13. Charles Foster, CMSR; findagrave.com/memorial/26510345/charles-foster; Charles Foster Biography, ETSU; *Diary of Charles Foster,* 66, Charles Foster Papers, TSLA.

14. William S. Fergusson biographical sketch and "War Memories," William S. Fergusson Papers, TSLA; William S. Fergusson, CMSR; Gallagher, ed., *Fighting for the Confederacy,* 307.

15. Fergusson, "War Memories," William S. Fergusson Papers, TSLA.

16. J. L. Tucker, CMSR; Fergusson, "War Memories," William W. Fergusson Papers, TSLA; civilwarshades.org/document/map-of-middle-tennessee/, accessed September 10, 2020. The Sayer map is at Vanderbilt University in Nashville.

17. Baker, "Something of Battle Field Maps," 63.

18. Shiman, "Engineering and Command," 96–98; *New York Herald,* June 23, 1863.

19. Campbell, "The Lost War Maps of the Confederates," 429–81.

20. Ibid.; Ralph W. Stephenson, "Confederate Mapping," loc.gov/collections/civil-war-maps/articles-and-essays/history-of-mapping-the-civil-war/confederate-mapping/.

21. Andrew Jackson Riddle Biography and Buchanan to Glenn, September 18, 1864, A. J. Riddle Collection, University of Alabama; Nichols, *Confederate Engineers,* 88; Map of the Military Division of the West, November 1864, LC no. 2009579249.

8. We Want Engineers

1. Daniel, *Conquered,* 262–65; Haughton, *Training, Tactics, and Leadership,* 137–49; Newton, *Lost for the Cause,* 54–56, 235, 251–88; Bragg, *Joe Brown's Army,* 85.

2. Henry C. Force, John F. Steele, Edward B. Sayers, George H. Hazlehurst, and Thomas E. Marble, CMSR; findagrave.com/memorial/140973004/Edward-brydges-sayers; findagrave.com/memorial/61584900/thomas-elan-marble; hazlehurstga.gov//Colonel-Hazlehurst.aspx. Hess (*Fighting for Atlanta,* 20) also lists a Lieutenant G. H. Browne as having arrived in May, but I have been unable to find anything on this officer. I am counting in the nineteen figure the following: Leadbetter, Pickett, Helm, Hazlehurst, Coleman, Steele, Presstman, Green, Pharr, Gloster, Winston, Ramsey, Cobb, Clarkson, Davies, Prinz, Newcomb, Foster, and Margraves.

3. Houghton, *Training, Tactics, and Leadership,* 150–52; Hess, *Civil War in the West,* 222; McMurry, *Atlanta 1864,* 94.

4. *Memphis Appeal,* May 5, 6, 9, 1864; Davis, ed., *Diary of a Confederate Soldier,* 116; Wynne and Taylor, eds., *This War So Horrible,* 57–58.

5. Abstract from return of the Department of Alabama, Mississippi, and East Louisiana, March 1864, and abstract from return of the Army of Tennessee, April 1864, *OR,* 32 (3): 729, 866; Lockett to Rives, July 1, 1864, *OR,* 39 (2): 679; report of Orlando M. Poe, *OR,* 38 (1): 128–29. See also Shiman, "Sherman's Pioneers in the Campaign to Atlanta," 251–266.

6. Mackall to Cantey, May 10, 1864, *OR,* 38 (3): 687; Johnston, "Opposing Sherman's Advance to Atlanta," 266; abstract from tri-monthly return, May 10, 1864, and Loring to French, June 20, 1864, *OR,* 38 (4): 691, 782; Hess, *Fighting for Atlanta,* 36–38, 47; Wynne and Taylor, eds., *This War So Horrible,* 68; Taylor, *Orlando M. Poe: Civil War General and Great Lakes Engineer,* 155–56; Thienel, *Mr. Lincoln's Bridge Builders,* 188; *Chicago Tribune,* June 1, 1864.

7. Report of Joseph E. Johnston, *OR,* 38 (3): 615; Johnston, *Narrative of Military Operations,* 319–20; Fergusson, "War Memories," William W. Fergusson Papers, TSLA; Davis, *Texas Brigadier,* 140–41; Stephen Davis to Larry J. Daniel, April 20, 2021, author's collection; Hafendorfer, ed., *Civil War Journal of William L. Trask,* 144.

8. Hood, *Advance and Retreat,* 110–16.

9. Hess, *Fighting for Atlanta,* 54, 56–57. Hess supported Johnston's position that the ridge could have been held despite enfilade fire, although he quotes Orlando Poe, Sherman's chief engineer, as saying that the position "could have been easily turned." Hess is also critical of Morris, who "had more experience at drawing maps than in combat engineering."

10. Beaur, ed., *Soldiering,* 113. My thanks to Stephen Davis for calling this to my attention.

11. Lash, *Destroyer of the Iron Horse,* 144; *Memphis Appeal,* May 23, 24, 27, 1864; *Montgomery [Ala.] Weekly Advertiser,* May 30, 1864.

12. *Memphis Appeal,* May 25, 1864; McMurry, *Atlanta 1864,* 85–92; Tower, ed., *A Carolinian Goes to War,* 189; French, *Two Wars,* 200.

13. McMurry, *Atlanta 1864,* 92–93; Tower, ed., *A Carolinian Goes to* War, 188; Wynne and Taylor, eds., *This War So Horrible,* 79–87.

14. French to Polk, June 2, 1864, French to Jack, June 8, 1864, Jack to French, June 17, 1864, *OR,* 38 (4): 755, 764, 778; French, *Two Wars,* 199, 201.

15. Davis, *Texas Brigadier,* 183; Hess, *Kennesaw Mountain,* 51–54; Castel, *Decision in the West,* 285.

16. Wiley, ed., *Four Years on the Firing Line,* 296; Davis, *Texas Brigadier,* 193; Hess, *Kennesaw Mountain,* 57; Hess, *Fighting for Atlanta,* 99–100, 106.

17. McDonough, *William Tecumseh Sherman,* 506, 509–10; *Chicago Tribune,* July 16, 1864; *Delaware [Ohio] Gazette,* July 29, 1864. An archeological examination of the Confederate fortifications up to the Kennesaw Line revealed a preference for lunettes, redans, and redoubts. See Fryman, "Fortifying the Landscape," 47.

18. Green to mother, June 28, 1864, Green Family Papers, UVA.

19. Wynne and Taylor, eds., *This War So Horrible,* 99–100.

20. Hess, *Fighting for Atlanta,* 137–38; Tower, ed., *A Carolinian Goes to War,* 194; *Southern Confederacy* [Atlanta, Ga.], July 9, 1864; report of Orlando M. Poe, *OR* 38 (1): 129; White and Runion, *Great Things Are Expected of Us,* 127; Morris to Wintter, July 4, 1864, D. Wintter, CMSR.

21. Shoup, "Dalton Campaign," 262–65; Hess, *Fighting for Atlanta,* 137–40; Scaife, "The Chattahoochee River Line," 42–58; Shaffer, "The Chattahoochee River Line Today," 133–35; report of Orland Poe, *OR* 38 (1): 129–30; Symonds, *Joseph E. Johnston,* 317. My thanks to Michael Shaffer to informing me that Shoup's friends, and the general himself, always used "Frank" in addressing him. Michael found this information in the Shoup Papers at the University of the South in Suwanee, Tennessee.

22. *Chattanooga [Tenn.] Daily Rebel,* September 7, 1864; *Memphis Appeal,* March 29, 1864; Memoranda, July 9, 1864, *OR,* 38 (5): 872–73.

23. Witherspoon to Rowley, July 17, 27, 1864, horsesoldier.com/images/products/documents-and-paper-goods/letters/17516.

24. Hood, *Advance and Retreat,* 173–74; Hess, *Fighting for Atlanta,* 166, 168; Castel, *Decision in the West,* 388.

25. Taylor, *Orlando M. Poe: Civil War General and Great Lakes Engineer,* 181.

26. Danville Leadbetter, CMSR; Davis, *Texas Brigadier,* 303; Davis, "Full-Throated Defender of Hood," 16–17; *Morning News* (Savannah, Ga.), July 13, 1888; M. M. Farrow, CMSR; D. W. Currie, CMSR; Presstman to Rowley, July 27, 1864, horsesoldier.com/images/product/17/56503.jpg.

27. [Frobel], "The Georgia Campaign," 241–46.

28. Walter J. Morris, Wilbur F. Foster, John D. Thomas, George A. Hazlehurst, all in CMSR; Hess, *Fighting for Atlanta,* 95, 100; Shoup to Brown, July 29, 1864, and Hood to Bragg, August 1, 1864, *OR,* 38 (5): 930, 937; *List of Staff Officers of the Confederate States Army,* 132. Porter is shown in *OR* correspondence as "J. A. Porter." The index shows his full name as "John A. Porter." His actual name was Joshua A. Porter.

29. Shoup to Hardee, August 13, 1864, *OR*, 38 (5): 961; Davis, *Texas Brigadier*, 409-10; Lane, ed., "Diary," in *Times That Prove People's Principles*, 200-201.

30. Lane, ed., "Diary," in *Times That Prove People's Principles*, 200-201; report of Brigadier General Hiram B. Granbury, August 31, 1864, *OR*, 38 (2): 743; Castel, *Decision in the West*, 503.

31. Fergusson, "War Memories," William W. Fergusson Papers, TSLA.

32. [Frobel], "The Georgia Campaign," 246-47; Lane, ed., "Diary," in *Times That Prove People's Principles*, 203.

33. Wynne and Taylor, eds., *This War So Horrible*, 108.

9. The Pontoniers

1. Sword, *Embrace an Angry Wind*, 45-50.

2. *Memphis Appeal*, as quoted in *Montgomery [Ala.] Daily Advertiser*, November 9, 1864; Roman, *Military Operations of General Beauregard*, 2:283; Davis, *Into Tennessee and Failure*, 28; Shoup to Conscript Authorities, July 29, 1864, *OR*, 38, pt. 4, 930; abstract from return of the Army of Tennessee, August 31, 1864, *OR*, 45 (1): 520; Iobst, *Civil War Macon*, 341. The list in table 9.1 is a compilation of all available records. During the Tennessee Campaign, Hood and his corps commanders did not mention staff members, as was the earlier custom, thus leaving gaps in the record.

On September 18, 1864, Orland Poe entered the Confederate lines at Rough and Ready, Georgia, in a letter exchange. He mentions meeting Lieutenant Colonel William Proctor Smith, an engineer in the Army of Northern Virginia. Precisely what the Virginian was doing in Georgia at this late date is not known, but there is no indication that he had any connection with the Army of Tennessee. See Taylor, ed., *My Dear Nellie*, 264.

3. Daniel, *Conquered*, 313-15; Hess, *Civil War Supply and Strategy*, 180.

4. Robinson report, October 19, 1864, *OR*, 39 (3): 836; Lockett to Alexander, November 20, 1864, *OR*, 45 (1): 1220.

5. Hess, *Civil War Supply and Strategy*, 188; Roane to wife, October 21, 1864, Thomas Roane Letters, Mississippi Valley Collection; Roman, *Military Operations of General Beauregard*, 2:291, 293; Daniel, *Conquered*, 316-17.

6. Lee to Thomas, January 4, 1865, *OR*, 45 (2): 507; Davis, *Into Tennessee and Failure*, 100; Lindsay, "Capture of Florence, Ala., Under Hood," 423.

7. Talcott, "Reminiscences of the Confederate Engineer Service," 260; Partridge, *History of the Ninety-Sixth Regiment*, 630-31; Duane, *Manual for Engineer Troops*, 17-18; Billings, *Hard Tack and Coffee*, 384-88.

8. Gibson to Lawton, April 11, 1864, *OR* 32 (3): 772 (120 boats); Wilder to Garfield, November 8, 1864, August 22, 1863, *OR*, 30 (3): 122 (47 boats); Memorandum for Colonel Brown, February 8, 1863, *OR*, 32 (2): 679 (130 wagons); James Lanning Diary, October 31, 1864, Bell Wiley Collection, EU; Traynham to wife, November 13, 1864, M. A. Traynham Letter, Confederate Collection, EU; Lindsay, "Capture of Florence, Ala., Under Hood," 423; Williams to Bullock, November 8, 1864, Thomas to Halleck, November 11, 1864, and Forrest to Taylor October 12, 1864, *OR* 39 (3): 902 (reserve pontoons), 746 (piers) 816 (piers); Lee to Thomas, December 27, 1864, *OR*, 45, pt. 2, 371 (dozen barges); Hughes, *Stephenson Memoirs*, 274.

9. Brown, ed., *One of Cleburne's Command*, 144; Sheppard, *By the Noble Daring of Her Sons*, 205; White and Runion, eds., *Great Things Are Expected of Us*, 143a; Thatcher, *A Hundred Battles*, 350-53.

10. M. A. Traynham to wife, November 13, 1864, M. A. Traynham Letter, Confederate Collection, EU.

11. Jones and Martin, eds., *The Gentle Rebel,* 110; Cate, ed., *Two Soldiers,* 149; Hamilton to mother, November 13, 1864, Joslyn, ed., *Charlotte's Boys,* 286; Journal of Brigadier General Francis A. Shoup and Report of Lieutenant General Stephen D. Lee, *OR,* 39 (1): 808, 811; Mason to Brent, November 9, 1864, *OR,* 39 (3): 904; Itinerary of Cheatham's Corps, *OR,* 45 (1): 730; Green to mother, November 13, 1864, Green Family Papers, UVA; *Chicago Tribune,* November 16, 1864; Brent to Presstman, November 2, 1864, *OR,* 52 (2): 772-73.

12. Beauregard to Cooper, October 24, 1864, *OR,* 39 (1): 796; Taylor to Grant, October 23, 1864, Taylor to Brent, October 27, 1864, and Ross to Roddey, October 28, 1864, *OR,* 39 (3): 845, 855-56, 864; Buford to Smith, November 15, 1864, *OR,* 45 (1): 1212; Report of J. Palmer, January 10, 1865, *OR* 45 (1): 642 (partial repair of Memphis & Charleston); Fleming to Sims, August 8, 1864 (twenty-seven carloads), Mobile & Ohio Railroad, csa-railroads.com (website by David L. Bright).

13. Gardner to Sturget, November 11, 1864, *OR,* 39 (3): 912.

14. Lockett to Alexander, December 20, 1864, *OR,* 45 (2): 716-17; Sword, *Shiloh,* 43.

15. Lockett to Alexander, December 20, 1864, *OR,* 45 (2): 716-17.

16. White and Runion, eds., *Great Things Are Expected of Us,* 144-45; McMurry, *John Bell Hood,* 169-70; Sword, *Embrace an Angry Wind,* 98. One postwar account claims that some of the pontoons were drawn by Texas longhorn steers, but other accounts state oxen. The latter seems more plausible.

17. Hess, *Civil War Supply and Strategy,* 215-21, 230-31; Sword, *Embrace an Angry Wind,* 412, 415; "Captains R. L. Cobb and F. P. Gracey," 249; Smith, *In the Lion's Mouth,* 206; Headquarters, Walthall's Division, January 3, 1865, *OR,* 45 (1): 728; Wilson to Whipple, December 25, 1864, and Mason to Stewart, December 23, 1864, *OR,* 45 (2): 351, 726.

18. Journal of the Army of Tennessee, December 25, 1864, and Headquarters, Walthall's Division, January 3, 1865, and Mason to Stewart, December 24, 1864 (three dispatches), *OR,* 45 (2), 507, 729-30; "Captains R. L. Cobb and F. P. Gracey," 249; McMurry, *John Bell Hood,* 181.

19. Itinerary of Cheatham's Army Corps, *OR,* 45 (1), 732; Lee to Welles, December 27, 1864, and to Thomas, January 4, 1865, *OR,* 45 (2), 371, 507.

20. Goodwin, "Capture of Hood's Supply and Pontoon Train," 483-84; Report of Colonel William J. Palmer, *OR,* 45 (1), 642-43; Beauregard to Cooper, January 22, 1865, *OR,* 45 (2), 804; Kirk, *History of the Fifteenth Pennsylvania Volunteer Cavalry,* 447-48, 458-59. Hood would blame the loss of the train on Roddey's cavalry brigade, which largely dispersed after it departed Decatur. See Reports of General John B. Hood, *OR,* 45 (1), 655-56.

21. Baker, "Something of Battlefield Maps," 63; Kundahl, *Confederate Engineer,* 264.

Epilogue

1. Findagrave.com/memorial/1540721/George-merideth-helm; Geary, ed., *Celine,* 243-44.

2. wgfd.wyo.gov/Get-Involved/Outdoor-Hall-of-Fame/Colonel-William D.-Pickett; *Baltimore Sun,* September 27, 1902; *The Citizen* (Frederick, Md.), October 3, 1902.

3. *Morning News* (Savannah, Ga.), December 23, 1883.

4. Gilmer to Von Sheliha, August 24, 1864, Gilder Lehrman Collection, GLC00908; "Brief History of Lemuel P. Grant," Atlanta Preservation Center; Allardice, *Confederate Colonels,* 242.

5. *Morning News* (Savannah, Ga.), July 13, 1888; findagrave.com/memorial/40666768/busrod-washington-frobel.

6. findagrave.com/memorial/85768745/henry-newton-pharr; "Engineer for the Army of Tennessee," 148; findagrave.com/memorial/43102641/Arthur-willis-gloster; reigelridge.com/roots/index.htm?ssmain=p426.htm; Creighton and Quarles, *Life of William Fisk Foster,* 50–58.

Appendix A

1. Solonick, *Engineering Victory,* 228.

Bibliography

Manuscripts

Arizona Historical Society, Tucson, Arizona
    Andrew Belcher Gray Biographical Summary
Atlanta Preservation Center, Atlanta, Georgia
    "Brief History of Lemuel P. Grant"
Auburn University, Auburn, Alabama
    Richard McCalla Letters
Benjamin F. Cooling Collection, Washington, D.C.
    John Hayden Journal
Duke University, Durham, North Carolina
    David B. Harris Papers
East Tennessee State University, Johnson City, Tennessee
    Charles Foster Biography
Emory University, Atlanta, Georgia
    Daniel L. Kelly Letters
    James Lanning Diary, Bell Wiley Collection
    M. A. Traynham Letter
Gilder Lehrman Collection, Chestertown, Maryland
    Corinth Fortifications, GLC05668
    Gilmer to Von Sheliha, August 24, 1864, GLC00908
Kennesaw Mountain National Battlefield Park, Kennesaw, Georgia
    "Condition of Pickett's Company of Sappers and Miners," MC 28
Louisiana State University, Baton Rouge, Louisiana
    "Confederate Topographical Report of Port Hudson"
Memphis Public Library, Memphis, Tennessee
    Gideon J. Pillow Papers
National Archives, Washington, D.C.
    Compiled Military Service Records
    Confederate Engineer Department, Letters and Telegrams Sent, 1861–1864, chap. III, vol. 3, National Archives Record Group 109
    Confederate Engineer Department, Letters and Telegrams Sent, Chief Engineer, Western Department, 1861–62, chap. III, vol. 8, National Archives Record Group 109

BIBLIOGRAPHY

Doug Schanz Collection, Roanoke, Virginia
    William D. Pickett Report, December 5, 1863
Southern Historical Collection, University of North Carolina, Chapel Hill, North Carolina
    Jeremy F. Gilmer Papers
    James I. Hall Letters
    Samuel H. Lockett Papers
Tennessee State Library and Archives, Nashville, Tennessee
    William Fergusson, "War Memories"
    Albert Fielder Diary
    Isham Harris Papers
Tulane University, New Orleans, Louisiana
    Henry Ginder Papers, Louisiana Research Center (LaRC)
    "Military Biography of Leon Joseph Fremaux," in Leon Joseph Fremaux Papers
University of Kentucky Library, Lexington, Kentucky
    "Biographical Sketch of William D. Pickett"
University of Virginia, Charlottesville, Virginia
    Green Family Papers, MSS 6211
Western Reserve Historical Society, Cleveland, Ohio
    Braxton Bragg Papers

Newspapers

Alexandria [Va.] Gazette
Athens [Tenn.] Post
Augusta [Ga.] Weekly Constitutionalist
Chattanooga [Tenn.] Daily Rebel
Chicago Tribune
Clarksville [Tenn.] Chronicle
Clarksville [Tenn.] Jeffesonian
Columbia [Tenn.] Herald
Forrest City [Ark.] Times
Memphis Appeal
New Orleans Delta
New Orleans Picayune
New York Herald
New York Times
Richmond [Va.] Daily Dispatch
St. Louis Republican
The [Savannah, Ga.] Morning News

Published Primary Sources

Abbot, Henry L. *Memoir of Dennis Hart Mahan, 1802–1871, Read before the National Academy, November 7, 1878.* Washington, D.C.: Judd and Detweiler, Printers, n. d.

Abrams, Alex St. Clair. "A Full and Detailed History of the Siege of Vicksburg." Atlanta: Intelligencer Steam Power Presses, 1863.

A. C. C. "The Dursting of 'Lady Polk.'" *Confederate Veteran* 12 (1904): 118-19.

Baker, Alpheus. "Island No. 10." *Southern Bivouac* (1883-84): 54-62.

Baker, Charles F. "Something of Battle Field Maps." *Confederate Veteran* 21 (February 1913): 62-63.

Billings, John D. *Hard Tack and Coffee*. Boston: G. M. Smith and Company, 1887.

Blackford, W. W. *War Years with Jeb Stuart*. New York: Charles Scribner's Sons, 1946.

Boyd, William K. *Military Reminiscences of William R. Boggs*. Durham, N. C.: Seeman Printery, 1913.

Brown, Norman D., ed. *One of Cleburne's Command: The Civil War Reminiscences and Diary of Capt. Samuel T. Foster, Granbury's Texas Brigade, CSA*. Austin: Univ. of Texas Press, 1980.

[Buchanan, Andrew H]. "From Engineer for the Army of Tennessee." *Confederate Veteran* 14 (August 1906): 369-71.

Buell, Don Carlos. "Operations in North Alabama." In *Battles and Leaders of the Civil War*, edited by Robert U. Johnson and Clarence C. Buel, 2:701-08. New York: Thomas Yoseloff, 1956.

Campbell, Albert H. "The Lost War Maps of the Confederates." *Century Magazine* 35 (January 1888), 479-81.

Cate, Wirt A., ed. *Two Soldiers: The Campaign Diaries of Thomas K. Key, C S. A. and Robert J. Campbell, U. S. A*. Chapel Hill: Univ. of North Carolina Press, 1938.

Davis, William C. *Diary of a Confederate Soldier: John S. Jackman of the Orphan Brigade*. Columbia: Univ. of South Carolina Press, 1990.

Foster, Wilbur F. "Battle Field Maps in Georgia." *Confederate Veteran* 20 (1912): 369-70.

———. "Building of Forts Henry and Donelson." In *Battles and Sketches of the Army of Tennessee*, edited by Broomfield L. Ridley. Mexico, Mo.: Missouri Printing, 1906.

French, Samuel G. *Two Wars: An Autobiography of Gen. Samuel G. French*. 1901. Reprint, Huntington, W.Va.: Blue Acorn Press, 1999.

Frobel, B. W. "Diary." In *Times That Prove People's Principles: Georgia in War*, edited by Mills Lane, 199-204. Savannah, Ga.: Beehive Press, 1993.

[Frobel, Bushrod W.]. "The Georgia Campaign; or A South-Side View of Sherman's March to the Sea." *Scott's Magazine* 5 (May 1868), 241-46.

Fussell, Joseph H. "Narrative of Interesting Events Prior to the Battle of Shiloh." *Historical Maury* 11 (December 1965): 384-95.

Gallagher, Gary W., ed. *Fighting for the Confederacy: The Personal Recollections of General Edward Porter Alexander*. Chapel Hill: Univ. of North Carolina Press, 1989.

Geary, Patrick J., ed. *Celine: Remembering Louisiana 1850-1871*. Athens: Univ. of Georgia Press, 1987.

"General R. B. Snowden and Staff." *Confederate Veteran* 3 (June 1895): 186.

Godwin, A. J. "Capture of Hood's Supply and Pontoon Train." *Confederate Veteran* 26 (November 1918): 483-84.

Hallock, Judith Lee, ed. *The Civil War Letters of Joshua K. Callaway*. Athens: Univ. of Georgia Press, 1997.

Hogane, J. T. "Reminiscences of the Siege of Vicksburg." *Southern Historical Society Papers* 11 (July 1883): 291-97.

Hood, John Bell. *Advance and Retreat: Personal Experiences in the United States and Confederate States Armies*. Bloomington: Indiana Univ. Press, 1959.

Hughes Nathaniel C., ed. *The Civil War Memoir of Philip Daingerfield Stephenson, D. D.: Private, Company K, 13th Arkansas Volunteer Infantry, Leader, Piece No. 4, 5th Company Washington Artillery, CSA*. Conway, Ark.: UCA Press, 1995.

Johnston, Joseph E. "Jefferson Davis and the Mississippi Campaign." In *Battles and Leaders of the Civil War*, edited by Robert U. Underwood and Clarence C. Buel, 3:472–82. New York: Thomas Yoseloff, 1956.

———. *Narrative of Military Operations During the Civil War*. 1874. Reprint; New York: Da Capo Press, 1959.

———. "Opposing Sherman's Advance to Atlanta." In *Battles and Leaders of the Civil War*, edited by Robert U. Johnson and Clarence C. Buel, 4: 260–77. New York: Thomas Yoseloff, 1956.

Johnston, William Preston. *The Life of Gen. Albert Sidney Johnston, Embracing His Services in the Armies of the United States, the Republic of Texas, and the Confederate States*. New York: D. Appleton, 1879.

Joslyn, Maurice Phillips, ed. *Charlotte's Boys: Civil War Letters of the Branch Family of Savannah*. Berryhill, Va.: Rockbridge, 1996.

Kirk, Charles H., ed. *History of the Fifteenth Pennsylvania Volunteer Cavalry Which Was Recruited and Known as the Anderson Cavalry in the Rebellion of 1861–1865*. Philadelphia: Privately published, 1906.

Lindsay, R. H. "Capture of Florence, Ala., Under Hood." *Confederate Veteran* 12 (December 1896): 423.

Lockett, S. H. "The Defense of Vicksburg." In *Battles and Leaders of the Civil War*, edited by Robert U. Underwood and Clarence C. Buel, 3:482–92. New York: Thomas Yoseloff, 1956.

———. "Surprise and Withdrawal from Shiloh," in *Battles and Leaders of the Civil War*, edited by Robert U. Johnson and Clarence C. Buel, 1:604–6. New York: Thomas Yoseloff, 1956.

Longstreet, James. *From Manassas to Appomattox: Memoirs of the Civil War in America*. 1896. Reprint, Bloomington: Indiana Univ. Press, 1960.

Mahan, D. H. *A Treatise on Field Fortifications, Containing Instructions on the Methods of Laying Out Constructing, Defending, and Attacking Intrenchments; with the General Outlines also of the Arrangement, the Attack and Defense of Permanent Fortifications*. Richmond, Va.: West and Johnston, 1862.

Partridge, Charles, ed. *History of the Ninety-Sixth Regiment, Illinois Volunteer Infantry*. Chicago: Brown, Pettibone, 1887.

Pickett, William D. "The Bursting of 'Lady Polk.'" *Confederate Veteran* 12 (1904): 277–78.

———. *Sketch of the Military Career of William J. Hardee*. Lexington, Ky.: James E. Hughes, 1910.

"River Batteries at Fort Donelson." *Confederate Veteran* 4 (1896), 393–400.

Russell, William. *My Diary North and South*. Boston: T. O. H. P. Burnham, 1862.

Shoup, Francis A. "Dalton Campaign—Works at Chattahoochee River—Interesting History." *Confederate Veteran* 3 (September 1895): 262–65.

Sims, L. Moody, ed. "A Louisiana Engineer at the Siege of Vicksburg." *Louisiana History* 8 (1967), 371–78.

Skellie, Ron, ed. *Lest We Forget: The Immortal Seventh Mississippi*. 2 vols. Birmingham, Ala.: Banner Digital Printing and Publishing, 2012.

Sorrell, Gilbert Moxley. *Reflections of a Confederate Staff Officer*. Jackson, Tenn.: McCowat-Mercer, 1958.

Stevenson, William G. *Thirteen Months in the Rebel Army.* New York: Barnes, 1864.
Talcott, T. M. R. "Reminiscences of the Confederate Engineer Service." In *Photographic History of the Civil War,* edited by Francis T. Miller, 5:256–70. New York: Thomas Yoseloff, 1956.
Taylor, Jesse. "The Defense of Fort Henry." In *Battles and Leaders of the Civil War,* edited by Robert U. Johnson and Clarence C. Buel, 1:368–72. New York: Thomas Yoseloff, 1957.
Tower, R. Lockwood, ed. *A Carolinian Goes to War: The Civil War Narrative of Arthur Middleton Manigault; Brigadier General, C.S.A.* Columbia: Univ. of South Carolina Press, 1983.
Trautman, Frederick, ed. *A Prussian Observes the American Civil War: The Military Studies of Justus Scheibert.* Columbia, Mo.: Univ. of Missouri Press, 2001.
von Scheliha, Viktor, *A treatise on coast defense: based upon the experience gained by officers of the Corps of Engineers of the Army of the Confederate States.* London: E. & F. N. Spon, 1868.
Walke, Henry. "Gunboats at Belmont and Fort Henry." In *Battles and Leaders of the Civil War,* edited by Robert U. Johnson and Clarence C. Buel, 1:358–72. New York: Thomas Yoseloff, 1956.
———. "The Western Flotilla at Fort Donelson, Island Number Ten, Fort Pillow, and Memphis." In *Battles and Leaders of the Civil War,* edited by Robert U. Johnson and Clarence C. Buel, 1:430–52. New York: Thomas Yoseloff, 1956.
Wallace, Lew. "The Capture of Fort Donelson." In *Battles and Leaders of the Civil War,* edited by Robert U. Johnson and Clarence C. Buel, 1:398–428. New York: Thomas Yoseloff, 1956.
Welker, David A., ed. *A Keystone Rebel: The Civil War Diary of Joseph Garey, Hudson's Battery, Mississippi Volunteers.* Gettysburg, Pa.: Thomas Publications, 1996.
White, William L., and Charles D. Runion, eds. *Great Things Are Expected of Us: The Letters of Colonel C. Irvine Walker, 10th South Carolina, C. S. A.* Knoxville: Univ. of Tennessee Press, 2009.
Wiley, Bell Irvin, ed. *Four Years on the Firing Line.* Jackson, Tenn.: McCowat-Mercer, 1963.
Wynne, Lewis N., and Robert A. Taylor, eds. *This War So Horrible: The Civil War Diary of Hiram Smith Williams.* Tuscaloosa: Univ. of Alabama Press, 1993.

Official Documents

The Biographical Directory of the Railroad Officials of America for 1887. Chicago: Railway Age Publishing Company, 1885.
Gardner's New Orleans Directory for 1861. New Orleans: Charles Gardner, 1861.
List of Staff Officers of the Confederate States Army 1861–1865. Washington, D.C.: Government Printing Office, 1891.
Mitchell, John L. *Tennessee State Gazetteer and Business Directory for 1860–61.* Nashville: John L. Mitchell, 1860.
Rowland, Dunbar, ed. *Encyclopedia of Mississippi History: Comprising Sketches of Counties, Towns, Events, Institutions and Persons Arranged in Cyclopedic Form,* vol. 1. Madison, Wisc.: Selwyn A. Bryant, 1907.
US Census, 1860.
US Navy Department, *The Official Records of the Union and Confederate Navies.* 30 vols. Washington, D.C.: Government Printing Office, 1894–1922.
US War Department. *The Official Military Atlas of the Civil War.* Washington, D.C.: Government Printing Office, 1891–1895.

BIBLIOGRAPHY

———. *The War of the Rebellion: A Compilation of the Official Records of the Union and Confederate Armies.* 128 vols. Washington, D. C.: Government Printing Office, 1880–1901.

Williams City Directory for Memphis, 1860. Memphis, Tenn.: N.p., 1860.

Published Secondary Sources

Allardice, Bruce. *Confederate Colonels: A Biographical Register.* Columbia: Univ. of Missouri Press, 2008.

American Society of Civil Engineers. *Transactions of the American Society of Civil Engineers.* New York, 1896–1917.

"An Honored Veteran of Two Wars." *Confederate Veteran* 22 (May 1914), 207.

Army, Thomas F., Jr. *Engineering Victory: How Technology Won the Civil War.* Baltimore: John Hopkins Univ. Press, 2016.

Ballard, Michael B. *Pemberton: The General Who Lost Vicksburg.* Jackson: Univ. Press of Mississippi, 1991.

Bearss, Edwin C. *The Campaign for Vicksburg.* 3 vols. Dayton: Morningside, 1986.

Bearss, Edwin C., and Warren E. Grabau. "How Porter's Flotilla Ran the Gauntlet Past Vicksburg." *Civil War Times Illustrated* 1 (December 1962), 38–48.

Beers, Henry P. "A History of the U.S. Topographical Engineers, 1813–1863." *The Military Engineer* 34 (June 1942), 287–92; 34 (July 1942): 348–52.

Bragg, William Harris. *Joe Brown's Army: The Georgia State Line, 1862–1865.* Macon, Ga.: Mercer Univ. Press, 1987.

"Capt. J. K. P. McFall." *Confederate Veteran* 16 (December 1908): 656.

"Captains R. L. Cobb and F. P. Gracey." *Confederate Veteran* 3 (August 1895): 249.

Castel, Albert. *Decision in the West: The Atlanta Campaign of 1864.* Lawrence: Univ. Press of Kansas, 1992.

Cathey, M. Todd, and Gary W. Waddey. *"Forward My Brave Boys!": A History of the 11th Tennessee Volunteer Infantry C. S. A. 1861–1865.* Macon, Ga.: Mercer Univ. Press, 2016.

Chernow, Ron. *Grant.* New York: Penguin Press, 2017.

Connelly, Thomas L. *Army of the Heartland: The Army of Tennessee, 1861–62.* Baton Rouge: Louisiana State Univ. Press, 1976.

Cooling, Benjamin F. *Fort Donelson's Legacy: War and Society in Kentucky and Tennessee, 1862–1863.* Knoxville: Univ. of Tennessee Press, 1997.

———. *Forts Henry and Donelson: The Key to the Confederate Heartland.* Knoxville: Univ. of Tennessee Press, 1987.

Creighton, Wilbur F. *The Life of William Fisk Foster: A Civil Engineer, Confederate Soldier, Builder, Churchman and Freemason.* Nashville: Ambrose Print Co., 1961.

Crute, Joseph H., Jr. *Units of the Confederate States Army.* Shippensburg, Pa.: White Mane Publishing Company, 1991.

Cummings, Charles M. *Yankee Quaker, Confederate General: The Curious Career of Bushrod Rust Johnson.* Rutherford, N.J.: Fairleigh Dickinson Press, 1971.

Davis, Stephen. "Full-Throated Defender of Hood." *Civil War News* 42 (May 1916), 16–17.

———. *Into Tennessee and Failure: John Bell Hood.* Macon, Ga.: Mercer Univ. Press, 2020.

———. *Texas Brigadier to the Fall of Atlanta: John Bell Hood.* Macon, Ga.: Mercer Univ. Press, 2019.

BIBLIOGRAPHY

Daniel, Larry J. *Battle of Stones River: The Forgotten Conflict between the Confederate Army of Tennessee and the Union Army of the Cumberland.* Baton Rouge: Louisiana State Univ. Press, 2012.

———. *Conquered: Why the Army of Tennessee Failed.* Chapel Hill: Univ. of North Carolina Press, 2019.

Daniel, Larry J., and Lynn N. Bock. *Island No. 10: Struggle for the Mississippi Valley.* Tuscaloosa: Univ. of Alabama Press, 1996.

"Engineer for Army of Tennessee." *Confederate Veteran* 3 (May 1895), 148.

Faust, Patricia L., ed. *H Historical Times Illustrated Encyclopedia of the Civil War.* New York: Harper and Row, 1986.

Field, Ron. *American Civil War Fortifications: The Mississippi and River Forts.* Oxford, UK: Osprey Press, 2007.

Fryman, Robert J. "Fortifying the Landscape: An Archeological Study of Military Engineering and the Atlanta Campaign." In *Archeological Perspectives on the American Civil War,* edited by Clarence C. Geier and Stephen R. Potter, 43–55. Gainesville: Univ. Press of Florida, 2000.

Grabau, Warren E. *Ninety-Eight Days: A Geographer's View of the Vicksburg Campaign.* Knoxville: Univ. of Tennessee Press, 2000.

Hazlett, James C., Edwin Olmstead, and M. Hume Parks. *Field Artillery Weapons of the Civil War.* Newark: Univ. of Delaware Press, 1983.

Hess, Earl J. *The Civil War in the West: Victory and Defeat from the Appalachians to the Mississippi.* Chapel Hill: Univ. of North Carolina Press, 2012.

———. *Civil War Supply and Strategy: Feeding Men and Moving Armies.* Baton Rouge: Louisiana State Univ. Press, 2020.

———. *Field Armies and Fortifications in the Civil War: The Eastern Campaigns, 1861–1864.* Chapel Hill: Univ. of North Carolina Press, 2005.

———. *Fighting for Atlanta: Tactics, Terrain, and Trenches in the Civil War.* Chapel Hill: Univ. of North Carolina Press, 2018.

———. *Kennesaw Mountain: Sherman, Johnston, and the Atlanta Campaign.* Chapel Hill: Univ. of North Carolina Press, 2013.

———. *The Knoxville Campaign: Burnside and Longstreet in East Tennessee.* Knoxville: Univ. of Tennessee Press, 2012.

———. "Revitalizing Traditional Military History." In *Upon the Fields of Battle: Essays on the Military History of America's Civil War,* edited by Andrew S. Bledsoe and Andrew F. Lang, 20–42. Baton Rouge: Louisiana State Univ. Press, 2018.

———. *Storming Vicksburg: Grant, Pemberton, and the Battles of May 19–22, 1863.* Chapel Hill: Univ. of North Carolina Press, 2020.

Hoffman, John. *The Confederate Collapse of the Battle of Missionary Ridge: The Reports of James Patton Anderson and His Brigade Commanders.* Dayton, Ohio: Morningside, 1985.

Hoffman, Mark. *"My Brave Mechanics": The First Michigan Engineers and Their Civil War.* Detroit, Mich.: Wayne State Univ. Press, 2007.

Horn, Huston. *Leonidas Polk: Warrior Bishop of the Confederacy.* Lawrence: Univ. Press of Kansas, 2019.

Houghton, Andrew. *Training, Tactics, and Leadership in the Confederate Army of Tennessee.* London: Frank Cass, 2000.

Hughes, Nathaniel C., Jr. *The Battle of Belmont: Grant Strikes South.* Chapel Hill: Univ. of North Carolina Press, 1991.

———. *General William J. Hardee: Old Reliable.* Wilmington, N.C.: Broadfoot Publishing Company, 1987.

———. *The Pride of the Confederate Artillery: The Washington Artillery of the Army of Tennessee.* Baton Rouge: Louisiana State Univ. Press, 1997.

Hughes, Nathaniel C., Jr., and Roy P. Stonesifer Jr. *The Life and War of Gideon J. Pillow.* Chapel Hill: Univ. of North Carolina Press, 1993.

Iobst, Richard W. *Civil War Macon: The History of a Confederate City.* Macon, Ga.: Mercer Univ. Press, 1999.

Jackson, Harry L., and Ronald A. Ellis. *The First Regiment Engineer Troops, P. A. C. S.* Louisa, Va.: R. A. E. Design and Publishing, 1998.

Jankins, Robert D., Sr. "Saving Dalton's Battlefields." *Blue and Gray* 32 (2015): 18.

"John G. Kelly," *Confederate Veteran* 11 (August 1902), 370.

Johnston, William Preston. *The Life of Gen. Albert Sidney Johnston, Embracing His Services in the Armies of the United States, the Republic of Texas, and the Confederate States.* New York: D. Appleton, 1879.

Krick, Robert L. *Staff Officers in Gray: A Biographical Register of the Staff Officers in the Army of Northern Virginia.* Chapel Hill: Univ. of North Carolina Press, 2003.

Kundahl, George G. *Confederate Engineer: Training and Campaigning with John Morris Wampler.* Knoxville: Univ. of Tennessee Press, 2000.

Lash, Jeffrey N. "Anthony L. Maxwell, Jr.: A Yankee Bridge Builder for the Confederacy, 1862–1865." *Journal of Confederate History* 6 (fall 1990): 106–36.

———. *Destroyer of the Iron Horse: General Joseph E. Johnston and Confederate Rail Transport, 1861–1865.* Kent, Ohio: Kent State Univ. Press, 1991.

Lindsley, John B., ed. *Military Annals of Tennessee, Confederate* (Nashville, Tenn.: J. M. Lindsley, 1886.

"Matthew F. Maury," *Confederate Veteran* 28 (January 1920), 29.

McDonough, James Lee. *War in Kentucky: From Shiloh to Perryville.* Knoxville: Univ. of Tennessee Press, 1994.

———. *William Tecumseh Sherman: In the Service of My Country.* New York: Norton, 2016.

McMurry, Richard M. *Atlanta 1864: Last Chance for the Confederacy.* Lincoln: Univ. of Nebraska Press, 2000.

——— *The Fourth Battle of Winchester: Toward a New Civil War Paradigm.* Kent, Ohio: Kent State Univ. Press, 2002.

———. *John Bell Hood and the War for Southern Independence.* Lexington: Univ. Press of Kentucky, 1982.

———. *Two Great Rebel Armies: An Essay in Confederate Military History.* Chapel Hill: Univ. of North Carolina Press, 1989.

Murray, Williamson, and Wayne Wei-Siang Hsieh. *A Savage War: A Military History of the Civil War.* Princeton, N.J.: Princeton Univ. Press, 2016.

Nichols, James Lynn. *Confederate Engineers.* Tuscaloosa, Ala.: Confederate Publishing, 1957.

Noe, Kenneth W. *Perryville: This Grand Havoc of Battle.* Lexington: Univ. Press of Kentucky, 2001.

Partin, Robert. "The Civil War in East Tennessee as Reported by a Confederate Railroad Bridge Builder." *Tennessee Historical Quarterly* 22 (September 1963): 238–58.

Polk, William K. *Leonidas Polk: Bishop and General.* 2 vols. New York: Longmans, Green, 1913.

Powell, David A. *The Chickamauga Campaign.* 3 vols. El Dorado, Calif.: Savas Beatie, 2014–2016.

BIBLIOGRAPHY

Ramage, James A. *Rebel Raider: The Life of General John Hunt Morgan.* Lexington: Univ. of Kentucky Press, 1986.

Robertson, William Glenn. *River of Death: The Chickamauga Campaign.* Chapel Hill: Univ. of North Carolina Press, 2018.

Roland, Charles. "Albert Sidney Johnston and the Defense of the Confederate West." *Confederate Generals in the Western Theater: Classic Essays on America's Civil War,* edited by Lawrence L. Hewitt and Arthur W. Bergeron, 1:13-23. Knoxville: Univ. of Tennessee Press, 2010.

Roman, Alfred. *The Military Operations of General Beauregard in the War Between the States, 1861 to 1865; Including a Brief Personal Sketch and Narrative of his Service in the War with Mexico, 1846-48.* 2 vols. New York: Harper and Bros., 1883.

"Samuel H. Lockett." *Confederate Veteran* 6 (April 1898), 181.

Scaife, William R. "The Chattahoochee River Line: An American Maginot." *North and South* 42 (May 2016), 16-17.

Shaffer, Michael K. "The Chattahoochee River Line Today." In *A Long and Bloody Task,* by Stephen Davis, 133-35. El Dorato Hills, Calif.: Savas Beatie, 2016.

Shea, William L., and Terrence J. Winschel. *Vicksburg Is the Key: The Struggle for the Mississippi River.* Lincoln: Univ. of Nebraska Press, 2003.

Sheppard, Jonathan C. *By the Noble Daring of Her Sons: The Florida Brigade of the Army of Tennessee.* Tuscaloosa: Univ. of Alabama Press, 2012.

Shiman, Philip L. "Engineering and Command: The Case of William S. Rosecrans." In *The Art of Command in the Civil War,* edited by Steven E. Woodworth, 84-117. Lincoln: Univ. of Nebraska Press, 1998.

———. "Sherman's Pioneers in the Campaign to Atlanta." In *The Campaign for Atlanta and Sherman's March to the Sea,* edited by Theodore P. Savas and Davis A. Woodbury. Campbell, Calif.: Savas Woodbury Publishers, 1994.

Smith, Derek. *In the Lion's Mouth: Hood's Tragic Retreat From Nashville, 1864.* Mechanicsburg, Pa.: Stackpole Books, 2011.

Smith, Timothy B. *Corinth 1862: Siege, Battle, Occupation.* Lawrence: Univ. Press of Kansas, 2012.

———. *Grant Invades Tennessee: The 1862 Battles for Forts Henry and Donelson.* Lawrence: Univ. Press of Kansas, 2009.

———. *Shiloh: Conquer or Perish.* Lawrence: Univ. Press of Kansas, 2014.

———. *The Union Assaults at Vicksburg: Grant Attacks Pemberton, May 17-22, 1863.* Lawrence: Univ. Press of Kansas, 2020.

Solonick, Justin S. *Engineering Victory: The Union Siege of Vicksburg.* Carbondale: Southern Illinois Univ. Press, 2015.

Swint, Philip L., and D. H. Mohler. "Eugene F. Falconnet, Soldier, Engineer, Inventor." *Tennessee Historical Quarterly* 21 (September 1962): 219-34.

Sword, Wiley. *Embrace an Angry Wind: The Confederacy's Last Hurrah: Spring Hill, Franklin, and Nashville.* New York: Harper Collins, 1992.

——— *Shiloh: Bloody April.* 1974. Reprint, Dayton: Morningside, 2001.

Symonds, Craig L. *Joseph E. Johnston: A Civil War Biography.* New York: W. W. Norton, 1992.

Taylor, Paul. *Orlando M. Poe: Civil War General and Great Lakes Engineer.* Kent, Ohio: Kent State Univ. Press, 2009.

Thienel, Phillip M. *Mr. Lincoln's Bridge Builders: The Right Hand of American Genius*. Shippensburg, Pa.: White Mane Publishing Company, 2000.

Warner, Ezra T. *Generals in Gray: Lives of the Confederate Commanders*. Baton Rouge: Louisiana State Univ. Press, 1959.

Weinert, Richard P. *The Confederate Regular Army*. Shippensburg, Pa.: White Mane Publishing Company, 1991.

Welsh, Jack D. *Medical Histories of Confederate Generals*. Kent, Ohio: Kent State Univ. Press, 1995.

Williams, T. Harry. *P. G. T. Beauregard: Napoleon in Gray*. Reprint. Baton Rouge: Louisiana State Univ. Press, 2011.

Winschel, Terrence. *Triumph and Defeat: The Vicksburg Campaign*. Mason City, Iowa: Savas, 1999.

"With Cumberland University 49 Years." *Confederate Veteran* 19 (September 1911): 421.

Woodworth, Steven E. *Jefferson Davis and His Generals: The Failure of Confederate Command in the West*. Lawrence: Univ. of Kansas Press, 1990.

———. *This Great Struggle: America's Civil War*. New York: Rowman and Littlefield, 2011.

"Worthy Widow Who Deserves a Pension." *Confederate Veteran* 17 (July 1909): 318.

Index

Abrams, St. Clair, 91
Adairsville affair, 124-26
Alexander, Edward Porter, 104, 106, 107, 117
Allen, Henry W., 44
Allen, James H., 116
Anderson, Adna, 21, 22, 23
Armstrong, R. H., 100
Army of Mississippi, 66, 121, 123, 125, 130
Army of Northern Virginia, x, 53, 61, 65, 96, 98, 103, 112, 119, 121, 134
Army of the Cumberland, 54, 56, 93, 102, 113, 118, 123, 128
Army of the Mississippi, 41, 51, 52, 76
Army of the Ohio, 104, 123
Army of Tennessee, 52, 56, 57, 63, 68, 194, 109, 121, 138, 148
Army of the Tennessee, 41, 42, 81, 123
Army of the West, 47
Army, Thomas F., Jr., 93
Atlanta Campaign, 137-38; defenses of, 68-69; Kennesaw line, 128-31; Chattahoochee River line, 131-33; siege of, 136
Atlanta, Georgia, 46, 59, 67, 68, 107, 116
Artillery fire, direct, 10, 24, 39, 30; plunging, 4, 9; penetration depth of, 17, 29, 36142, 148; enfilade, 137
Autry, James, 72

Bainbridge, Alabama, 142, 148
Baker, Alpheus, 128
Baker, Charles H., 118
Ballard, Michael, 84

Barns, Arthur S., 60
Barron River, 32, 51
Bearss, Edwin C., 74
Beauregard, P. G. T., 11, 13, 32, 35, 40, 42, 43, 45, 46, 65, 78, 139, 141, 142, 144, 145, 146; theory of fortifications, 18, 22-23, 33, 71; critical of Gilmer, 29
Belmont, battle of, 10
Benjamin, Judah, 9, 40
Big Black River, 84, 85; battle of, 86
Birmingham narrows, 22
Blackford, William, 98
Blessing, P. J., 80, 90
Blue Mountain, Alabama, 67, 68
Boggs, William R., 49
Bowen, Achilles, 25
Bowen, John S., 76
Bowling Green, Kentucky, 11, 13, 25, 27, 28, 31, 32, 33, 35, 38, 39, 49
Bragg, Braxton, 18, 14, 42, 46, 58, 60, 61, 62, 63, 100, 103, 105, 106, 108, 109, 114; and Shiloh, 43-44; and Kentucky Campaign, 49-56; and Stones River, 54-55; and Tullahoma campaign, 62-63; and McLemore's Cove, 102-3; inability to understand maps, 117-18
Breckinridge, John C., 41, 43, 44, 45, 47, 52, 55
Bridgeport, Alabama, 41, 61, 62, 63, 96, 102, 114, 141
Bridges, 21, 26; repair of, xii, 58-60, 96, 105; demolition of, 40, 109, 127; duplicate, 68; Howe truss, 68, 97
Brower, G. C., 80

INDEX

Browne, G. H., 111
Buchanan, Andrew H., xi, 54, 63, 64, 116, 119, 124, 125
Buckner, Simon B., 32, 35, 65, 98, 102, 103, 106
Buell, Don Carlos, 31, 33, 35, 38, 40, 41, 42, 44, 46, 47, 48, 50, 51
Bull, Rice C., 127
Burnside, Ambrose, 104, 106

Campbell, Albert H., 112, 119
Canal, Island No. 10, 15, 71; Vicksburg, 71
Carter, S. P., 58
Cassville, Georgia, 125, 127
Champion Hill, battle of, 86
Charleston, Tennessee, 65, 106, 109
Chattahoochee River, xii, 68, 124, 133, 140
Chattanooga, Tennessee, 40, 41, 46, 48, 51, 52, 53, 58, 65, 68, 100, 101, 102, 103, 106, 114, 115, 143
Cheatham, Benjamin F., 115, 131, 134, 139, 144, 146, 148
Cheatham's Hill, 131
Chickamauga, battle of, 103, 113
Clarke, John W., 106, 107
Clarkson, A. W., 98, 99, 109, 143, 144
Clarksville, Tennessee, 21, 27–28, 31, 32, 33, 33, 38
Clayton, Henry, 116
Cleburne, Patrick, 5, 49, 50
Cobb, Robert L., 61, 65, 98, 99, 148, 151
Coleman, Thaddeus C., 101, 109, 122, 133, 137
Columbus, Kentucky, 7, 8–11, 13, 18, 21, 29
Connelly, Thomas L., ix, 30–32
Comstock, Cyrus, 82
Conway, J. J., 79, 80
Cooling, Benjamin F., 35
Cooper, Samuel, 8, 9
Corinth, Mississippi, 14, 18, 38, 40, 43, 44, 45 46, 144, 145, 146, 149
Couper, James M., xi, 66, 78, 80
Crittenden, George, 40
Cullom, George, 13
Cumberland Gap, 49, 58

Cumberland River, 22, 30, 31, 37, 38, 51; importance of, 21
Currie, D. W., 111, 135

Dabney, Frederick Y., 80
Dalton, Georgia, 109, 110, 121, 123
Darrow, Amos S., 63, 64
Davies, James J., 98, 100
Davis, Jefferson, 5, 8, 11, 13, 52, 54, 74, 84, 87, 134, 139
Decatur, Alabama, 40, 41, 53, 148
Demopolis, Alabama, 66, 67, 123, 140, 146
Department No. 2, 8
Department of Alabama, Mississippi, and East Louisiana, 77, 136, 139, 140, 142
Department of East Tennessee, 47, 60, 104
Department of Mississippi and North Louisiana, 80
De Russy, Lewis, 9, 10, 39
De Saulles, Arthur B., 14, 39, 45
De Veuve, Henry, 85
Dexter, Amory, 67
Dixon, Joseph, xi, 26–27, 28, 29, 31, 32, 36, 39, 150
Donellan, George, 67, 76, 78, 80, 88, 140
Donelson, Daniel, 22, 23, 98
Duck River, 55, 62, 146, 147
Ducktown Copper Mines, 57
Duffy, Frank M., 98

East Tennessee & Georgia Railroad, 58, 68
East Tennessee & Virginia Railroad, 58, 68, 104
Elk River, 47, 62, 63
Engineer Bureau (C.S.), Engineers (C.S.), xii; organization of, x; tools and equipment of, 6, 9, 15, 33, 63–65, 82, 88, 89, 99–100, 110; tasks of, 38, 57, 90, 95, 96; civil engineers, 93–94; in pre-Civil South, 94; shortages of, 94, 95, 96
Engineer troops (C.S.), 96–98, 123; First Regiment, 98; Second Regiment, 98; Third Regiment (*see* Third Engineer Regiment)

198

INDEX

Engineers (US), Corps of Engineers, xii, 2, 81, 95; Corps of Topographical Engineers, 2, 112; First Michigan Engineers, 47, 48, 54, 93, 102; Pioneer Brigade, 54; First Missouri Engineers, 123
Estill, Thomas L., 25, 39
Etowah Iron Works, 68
Etowah River, 127

Fairbanks, Jason M., 42
Falconnet, Eugene F., 94
Farragut, David, 71, 72, 73, 75
Farrow, Miles M., 135
Fay, Calvin, 3, 110
Fergusson, William W., 78, 102, 103, 116, 117, 124
Fleece, George B., 95
Fleming, L. J., 95
Flooding, x, 7, 14, 15, 22, 24, 34
Florence, Alabama, 21, 41, 142, 143, 144, 146, 148
Floyd, John, 35
Flynn, W. O., 90
Force, Henry C., 53, 63, 115, 116, 117, 118, 121, 124
Forrest, Nathan Bedford, 46, 47, 68, 102
Fort Buckner, 108
Fort Clark, 28
Fort Cleburne, 5, 113
Fort De Russy, 10
Fort Donelson, 12, 28, 31, 34, 38, 39; site selection, 22; weakness of, 22, 29, 35, 36–37; battle of, 36
Fort Garrett, 83
Fort Harris, 3, 4, 7, 11
Fort Heiman, 33
Fort Henry, 12, 28, 32, 37, 38, 31, 71; site selection, 22, 23–24; weakness of, 24, 26, 34–35; attacked, 34
Fort Hill, 82
Fort Levin, 29
Fort Pemberton, 78
Fort Pickering, 3
Fort Pillow, 6, 7, 8, 10, 11, 12, 14, 18, 19, 39, 41, 71
Fort Polk, 6
Fort Rains, 62, 63

Fort Sanders, 107, 108
Fort Terry, 29
Fort Thomas, 81
Fort Wright, 4, 7

Fortifications: abatis, 29, 62, 63, 87, 88, 131, 137; lunette, 4, 82, 83, 87, 88; cremailliere, 7; embrasure, 7, 36, 87, 138; bombproof, 9, 17, 76; redoubt, 10, 62, 69, 84, 85, 91, 127, 133; parapet, 17, 26, 29, 34, 69, 92, 131; redan, 6, 14, 15, 17, 82, 83, 84, 87, 90, 91, 102, 112; tete-de-pont, 32, 132; traverses, 137; improvised, 35, 108, 128; stockade, 82, 87; chevaux-de-frise, 137, 138
Foster, Charles, 39, 49, 60, 98, 100, 116, 117
Foster, Wilbur F., 21–22, 24, 25, 39, 60, 116, 124, 132, 136, 152–53
Fremaux, Leon, 42–43, 45, 47, 66, 67, 79, 80, 113, 151
French, Samuel G., 124, 129
Freret, William, 49, 50
Frobel, Bushrod W., 134–35, 136, 137, 139, 152
Fryman, Robert J., 69

Gallimand, Jules V., 18, 80, 81
Gaines Frank, 116, 124
Gardner, Frank, 145, 146
Gilip, Captain, 109
Gillooly, Francis, 80
Gilmer, Jeremy, 17, 34–36, 39, 43, 44, 47, 49, 52, 53, 54, 55, 62, 63, 64, 66, 68, 77, 95, 86, 152; arrives in West, 27; criticism of, 29, 30–32
Ginder, Henry, 66, 79, 80, 88, 89
Glenn, John W. 110, 119
Glenn, T. J., 29, 31
Gloster, Arthur W., 11, 65, 79, 80, 98, 99, 101, 144, 149, 152
Godwin, A. J., 149
Gooden, C. C., 100
Goodwin, J. J., 98
Gordon, George W., 94
Gordon, G. Wash, 147
Gorgas, Josiah, x

INDEX

Grabau, Warren E., 74, 75
Grand Gulf, Mississippi, 75
Grant, Lemuel P., 58, 60, 67–70, 110, 116, 122, 134, 140
Grant, Ulysses S., 8, 19, 35, 36, 40, 41, 42, 44, 47, 82, 84, 86, 87, 88
Gray, Andrew B., 5–7, 14, 15, 150
Gray, Asa. *See* Gray, Andrew B.
Green, John W., 46, 47, 48, 52, 61, 63, 64, 66, 98, 122, 131, 140, 144, 148
Green, R. C., 95
Grose, William, 103
Gutherz, Gottfried G., 125, 130

Halleck, Henry W., 27, 33, 38, 41
Haney, Joseph H., 116
Hansell, William A., 70
Hardee, William J., 8, 32, 33, 35, 40, 41, 44, 47, 48, 51, 60, 61, 62, 63, 64, 66, 101, 109, 121, 123, 125, 128, 129, 139
Harris, David B., 12, 15, 18, 19, 39, 47, 48, 49, 51, 52, 53, 71, 72, 74, 76, 150
Harris, Isham G., 2, 3, 21, 23, 24, 40
Hart, William T., 99
Hassell, Bentley D., 95
Haydon, John, 22, 25, 26, 30, 33, 34, 39, 59, 150–51
Haynes, Milton A., 36
Hazlehurst, George H., 110, 121, 122, 131, 136
Heartland, ix–x
Heiman, Adolphus, 26, 29
Helm, George M., 39, 43, 49, 61, 63, 64, 100, 102, 109, 116, 122, 151
Henry, Gustavus A., 24, 29
Herman, Valentine, 116, 124
Hess, Earl, ix, 88, 106, 123, 127, 147
Hill, D. H., 101
Hindman, Thomas, 103
Hiwassee River, 106
Hogane, James T., 74, 79, 80, 87
Holston River, xiii, 58, 59, 104, 106
Hood, John Bell, 109, 123, 124, 125, 127, 129, 133, 138, 139, 140, 141; and Tennessee Campaign, 146–48

Howe, William, 68
Humphreys, J. H., 4
Huntsville, Alabama, 41
Huston, Menefee, 99, 100
Hyett, D. H., 79, 80

Island No. 10, 5–7, 8, 10, 11, 12, 13, 17–18, 71

Jackson, Mississippi, 67, 77; battle of, 86
Johnson, Bushrod, 23, 24, 25, 26, 39, 99
Johnston, Albert Sidney, 8, 9, 17, 27, 28, 30, 31, 32, 33, 35, 38, 40–43
Johnston, Joseph E., 46, 61, 66, 74, 84, 86, 87, 109, 110, 117, 121, 122, 123, 125, 127, 128, 129, 131, 132, 150, 133, 138; theory of fortifications, 12–13, 82, 84; relieved of command, 134
Johnston, William Preston, 24, 30, 35, 62
Jones, W. A. C., 67, 125
Jonesboro, Georgia, battle of, 136, 139
Jordan, Thomas, 78

Kelly, David L., 102
Kelly, John G., 66, 80, 81
Kennesaw Mountain. *See* Atlanta Campaign
Kentucky Campaign, 46–52
Kentucky neutrality, 7, 8, 23
Knight, J. P., 120
Knoxville, Tennessee, 49, 52, 53, 58, 60, 65, 68, 102, 105, 114, 117
Knoxville, siege of, 106–7
Kundahl, George G., x

Labor, 3–4, 5, 7, 8, 10, 15, 17, 26, 33, 46, 47, 69, 89–90, 133
Lacey, W. R., 61
Leadbetter, Danville, x, 104, 105, 106, 109, 110, 122, 143
Lee, Robert E., 11, 45, 97
Lee, Samuel P., 148
Lee, Stephen D., 134, 143, 147
Line Port, Tennessee, 29
Lockett, Samuel H., xi, 42, 43–46, 49, 66, 72, 76, 77–83, 85–90, 129, 140, 142, 145, 146, 152, 153

INDEX

Longstreet, James, 103, 105, 106, 107
Lookout Mountain, 108, 117
Loring, W. W., 85
Loudon, Tennessee, 61, 65, 106, 107
Louisville and Nashville Railroad, 27
Lovell, Mansfield, 38, 71
Lynch, Montgomery, 4, 10, 11, 19, 25, 39

Mackall, William, 17
Macon, Georgia, 118, 139–40
Mahan, Dennis Hart, xi, 2, 29, 34, 78, 92, 110, 121, 135
Mallory, Lee, 120
Maney, George, 136
Manigault, Arthur M., 108, 128
Mann, John G., 11, 25, 39, 46, 60, 98
Maps, 43, 45; availability of, 112; by Conrad Meister, 114–15; photo reproduction, 119–20
Marble, Thomas E., 111
Margraves, George R., 50, 98
Marsh, Daniel, 28
Martin, Albert, 113
Maury, Matthew Fontaine, Sr., 60
Maxson, G. W., 48
Maxwell, Anthony L. & Son, 59, 96, 97, 104
Maxwell, Anthony L., 68, 96
McCalla, Richard C., 60, 65, 99, 103, 104, 105
McCook, Alexander, 55
McCown, John P., 14, 15
McCulloch, W. J., 118
McFall, James K. P., 45, 54, 63, 64, 65, 115, 116, 122, 124
McGuire, J. W., 116
McLarin, Arch, 110
McLemore's Cove, 102–3
McMahon, Edward, 3, 80
McPherson, James, 92, 123
McRee, George R., 111
McVernon, S. McD., 67, 80
Meister, Conrad, 49, 63, 114–115
Memphis & Charleston Railroad, 45, 46, 47, 53, 97, 144, 146
Memphis, Clarksville & Louisville Railroad, 27

Memphis, Tennessee, 2, 4, 11, 15, 27, 40, 41
Meridian, Mississippi, 67, 140, 143, 146
Meriwether, Minor, 3, 10, 11, 19, 39, 43, 45, 47, 66, 80
Meriwether, Niles, 95
Merrill, William E., 118
Miller, Joseph A., 3, 9, 11, 34, 39
Mill Springs, battle of, 33
Mines. *See* Vicksburg
Minor, Otey Henry, 100
Missionary Ridge, 106, 108, 109
Mississippi River, 19, 21, 40, 80, 84; significance of, 1; defenses along, 2–3, 7, 41, 71
Mobile, Alabama, 67, 46, 66, 140, 146, 152
Mobile and Ohio Railroad, 45, 143, 144, 146
Moncure, Thomas J., 106
Moore, John, 49
Moreno, Theodore, 70, 122
Morgan, John Hunt, 54, 63, 64
Morris, R. C., 95
Morris, Walter, J., 28, 39, 49, 63, 64, 100, 109, 113, 125, 127, 128, 129, 136
Morrison, James, 100
Morton, James St. Claire, 54
Munfordville, Kentucky, 51
Murfreesboro, Tennessee, 40, 47, 52, 53, 54, 114

Nashville, Tennessee, 21, 27, 30, 31, 35, 38, 40, 42, 48, 52, 93, 114
Nashville & Chattanooga Railroad, 47, 48, 62, 68, 96
Nashville & Decatur Railroad, 45
Navy, US, 10, 17, 18, 30, 34, 35, 36, 38, 41, 71–72, 73
Negley, James, 102
Newcomb, T. S., 48, 63, 65, 98, 100
New Madrid, Missouri, 5, 6, 13, 14
New Orleans, Louisiana, 13, 18, 38, 47, 49, 66, 71, 152
Nichols, James, ix
Nocquet, James, 39–40, 43, 52, 53, 55, 65, 102, 103, 104

INDEX

O'Hea, Richard A., 67
Oliver, John, 109, 110, 128, 131
Oostanaula River, 68, 102, 104, 123

Parker, Nathan H., 79
Palmer, John, 100
Pattison, Helms A., 41, 78, 79, 113
Pegram, John, 44–45, 47
Pemberton, John C., 75, 77, 82, 84, 90, 92
Percy, Robert, 98
Perkins, James B., 100
Perryville, battle of, 52, 114
Pharr, Henry N., 11, 39, 47, 49, 51, 52, 63, 64, 65, 99, 102, 107, 148, 152
Pickett, George B., 39, 40, 45, 48, 102, 116, 122
Pickett, William D., 2–3, 4, 5, 9, 10, 11, 25, 33, 39, 43, 109, 151
Pillow, Gideon, 2, 4, 7, 8, 17, 35
Pine Bluff, Tennessee, 23
Pioneer Battalion (C.S.), 45
Pioneer Brigade. *See* Engineers (US), Corps of Engineers
Pittsburg Landing, Tennessee, 41, 43
Poe, Orlando, 124, 133
Polk, Leonidas, 5, 8, 9, 10, 12, 13, 21, 25, 26, 28, 31, 32, 33, 35, 40, 41, 44, 48, 51, 55, 56, 60, 63, 66, 67, 99, 113, 121, 123, 127, 128, 129
Pontoons, 54, 96, 100, 101, 114, 124, 127, 131, 133, 140; dimensions of, 53; crossing Tennessee River with, 102, 142–43, 148; construction of, 110, 146; McClellan style, 142; Hood's pontoon train destroyed, 149
Pope, John, 14, 40, 44, 153
Porter, David, 72, 73
Porter, Joshua A., 66, 67, 125, 136
Port Hudson, Louisiana, 71, 77, 79, 80, 81
Presstman's Engineer Battalion, 98
Presstman, Stephen W., 46, 47, 49, 61, 62, 63, 68, 98, 101, 103, 110, 115, 119, 122, 123, 129, 134, 136, 138, 140, 148, 150
Price, Sterling, 47
Prime, Frederick, 82, 93
Prinz, William D., 98, 100
Pritchard, M. B., 95

Ramsey, Waightetill A., 60, 99
Raven, T. G., 110
Resaca, battle of, 123
Richmond, Kentucky, battle of, 50–51
Riddely, R., 99, 100
Riddle, Andrew Jackson, 119
Ridley, Thomas Jefferson, 99, 100
River obstructions, 39; chains, 3, 4, 10; pilings, 5; sunken barges, 29
Rives, Alfred R., x, 110
Roberts, B. F., 110
Robinson, John M., 60, 105
Robinson, Powhatton, 66, 78–79, 80, 142
Roddey, Phillip, 148
Rogers, Charles C., 26
Roman, Alfred, 18
Rosecrans, William S., 54, 100, 102, 103
Rowley, Robert P., 7, 39, 53, 63, 102, 104, 133
Rucker, Edmund W., 3, 8, 25
Ruger, Edward, 113, 124
Russell, William, 3, 4

Sandbags, 34, 36, 88, 138
Sappers and miners, 3; Wintter's, 2–3, 4, 5, 11, 14–15, 18, 82; Gillimard's, 18, 81; Pickett's, 45, 46, 49, 99, 100, 102; Maxson's 49; Winston's, 49, 50, 99, 100; Jones's, 67; Porter's, 67; Clarkson's, 98
Saunders, John E., 33
Sayers, Edward B., 28, 30, 39, 49, 51, 63, 64, 100, 103, 118
Scheibert, Justus, 1, 6, 15
Schofield, John M., 123, 146
Seddon, James, 97
Selma, Alabama, 68, 140
Semmes, P. W., 99, 100
Shackleford, James M., 104
Shaffer, Michael, 132
Shelbyville, Tennessee, 60, 62, 100, 114, 118
Sheridan, Philip, 114
Sherman, William T., 67, 81, 87, 123, 127, 131, 128, 140, 133, 134, 138
Shiloh, battle of, 19, 43–44, 45, 113
Shoup, Francis, 87, 132

INDEX

Shoupades, 132
Smith, Ashbel, 88
Smith, Edmund Kirby, 47, 49–51, 57, 60
Smith, Felix R. R., 25, 27, 39, 60, 65, 98, 116, 131
Smith, Martin Luther, 72–74, 82, 87, 134, 135, 136, 139, 144, 145
Smith, Preston, 49
Smith, Timothy B., 71
Snowden, J. Hudson, 7, 10, 39
Solonick, Justin, 81, 92
Southard, R. R., 79, 80
Steele, John F., 63, 64, 116
Steele, Silvanus W., 42, 43, 54, 55, 63, 115, 116, 117, 118, 122
Stevenson, Alabama, 47, 114, 141
Stevenson, Carter, 54, 60
Stevenson, V. K., 30
Stewart, Alexander P., 18, 128, 134, 148
Stewart, John S., 116
Stockton, Philip, 2, 4
Strawberry Plains, Tennessee, 58, 59, 60, 61, 68
Stewart, Alexander P., 136
Stones River, battle of, 54–55, 113
Streight, Abel, 68
Square Fort, 83, 92

Taylor, Jesse, 24, 25, 34, 139
Taylor, Richard, 139
Tennessee Provisional Army, engineer corps, 23, 25
Tennessee River, 22, 31, 32, 41, 47, 63; importance of, 21
Third Engineer Regiment (C.S.), 95; staff of, 98; officers of, 100; Company A, 98, 105; Company B, 99, 102, 103, 107, 123; Company C, 98, 99, 102, 103, 110, 123; Company D, 106, 109, 110, 123; Company F, 99, 109, 123, 143; Company G, 98, 99, 109, 110, 123; Company H, 98, 99, 109, 123
Thomas, George, 123, 147, 148
Thomas, James D., 39, 49, 53, 65, 102, 116, 124, 136
Thysseus, Francois I. J., 78, 101

Tilghman, Lloyd, 31, 33, 34, 35
Tombigbee River, 46, 67, 140
Trask, William, 124
Traynham, M. A., 143, 144
Trezevant, J. T., 9
Trudeau, James, 14
Tucker, J. Louis, 116, 117
Tullahoma, Tennessee, 52, 55, 60, 61, 62, 100, 116, 118
Tupelo, Mississippi, 45, 46
Tyner, John S., 116, 122

US Corps of Engineers. *See* Engineers (US), Corps of Engineers
US Corps of Topographical Engineers. *See* Engineers (US), Corps of Engineers

Van Dorn, Earl, 44, 62, 78
Van Leer, Rush, 49
Vicksburg, Mississippi, 54, 71; river defenses of, 72, 74–75; soil, 74, 89; land defenses of, 82–84, 87; May 19–22 assaults, 87–88; siege of, 88–90; criticism of defenses, 91–92
Vernon, McD., 145
Villepique, John, 18
Vinet, John Baptist, 66, 125, 129
Von Sheliha, Victor, 17, 18, 39, 141, 142, 152

Walker, Irvine C., 100, 146
Walker, Leroy P., 2
Walker, W. H. T., 129
Walker, Marsh, 9
Wampler, John M., 42, 45, 46, 49, 54, 61, 62, 63, 64, 65, 100, 150
Wardly & Pikes, 60
Watauga River, 58, 104, 103
Watts, G. O., 31
Westbrook, J. L., 98
Western & Atlanta Railroad, 67, 68, 102, 110, 116, 123, 132
Western Military Institute, 23, 111
West Point, as engineering school, xi, 2
Wheeler, Joseph, 116, 123, 124
Whiting, W. H. C., x

203

Williams, E. P., 98
Williams, Hiram, 98
Williams, James T., 60
Wilson, James H., 146, 147, 148
Winschel, Terry, 83
Winchester, Napoleon B., 60, 116, 124
Winston, Edmund, 49, 50, 99, 133, 139
Wintter, Deitrich, 3, 80, 90, 125, 132, 140, 142, 146

Witherspoon, John N., 133
Wood, Thomas, 147
Woodworth, Stephen, 1
Wrenshall, John C., 53, 116, 118, 119, 122

Yazoo River, 71, 76, 78, 79, 81, 84, 86, 87, 140

Zollicoffer, Tennessee, 58, 61, 104